North American Mixing Forum 2023

Held at the 2023 AIChE Annual Meeting

Orlando, Florida, USA
5-10 November 2023

ISBN: 978-1-7138-9292-2

Printed from e-media with permission by:

Curran Associates, Inc.
57 Morehouse Lane
Red Hook, NY 12571

proceedings.com

Some format issues inherent in the e-media version may also appear in this print version.

AIChE
120 Wall Street, FL 23
New York, NY 10005-4020

Phone: (800) 242-4363
Fax: (203) 775-5177

www.aiche.org

Additional copies of this publication are available from:

Curran Associates, Inc.
57 Morehouse Lane
Red Hook, NY 12571 USA
Phone: 845-758-0400
Fax: 845-758-2633
Email: curran@proceedings.com
Web: www.proceedings.com

TABLE OF CONTENTS

DIVISION PLENARY: NORTH AMERICAN MIXING FORUM AWARDS SESSION (INVITED TALKS)

EXPERIMENTAL INVESTIGATIONS OF MIXING PROCESSES

MIXING IN MULTIPHASE SYSTEMS: COMPUTATIONAL MODELING

MIXING SCALE-UP AND SCALE-DOWN ISSUES IN PHARMACEUTICAL AND BIOPHARMACEUTICAL PROCESSES

MIXING IN MULTIPHASE SYSTEMS: EXPERIMENTS AND VALIDATION OF COMPUTATIONAL MODELS

192a On the Trailing Vortices behind Impeller Blades in Stirred Vessels

Harry Van den Akker

A still increasing computational speed and power, among other things due to the use of graphics processing units (GPUs), reveals ever more details of the flow fields in, among other devices, stirred vessels. A decade or more ago, we were happy to use vorticity to show that our simulations were capable of resolving the trailing vortices behind impeller blades, at least qualitatively. These days, there is a growing tendency to try and predict these trailing vortices quantitatively, and to use then these data for forecasting particle trajectories and bubble or droplet breakup in the impeller domain. Time for a review of the current state of the art in computationally simulating particularly these trailing vortices.

Over the years, both experimental and computational techniques have developed tremendously. We now have Stereo Particle Tracking Velocity with 2 or 4 high-speed cameras on the experimental side alongside Large Eddy Simulations (based on the lattice Boltzmann technique) and Detached Eddy Simulations, often supplemented with Lagrangian particle tracking routines, which are increasingly replacing the common RANS-based simulation techniques (on the basis of k-epsilon modelling). Of course, this leaves the option open that old experimental as well as old calculated data get outdated. However, what both the novel experimental and simulation techniques have in common is that they heavily rely on numerical techniques including data processing routines which are outside the scope of most users.

In this review, older and recent data about the flow over impeller blades from both experiments and simulations are evaluated and mutually assessed. The focus is on the trailing vortices behind impeller blades of both Rushton and Pitched-Blade impellers. The intriguing question is addressed whether or not, in these trailing vortices or otherwise, velocity vector magnitudes in excess of the impeller tip speed are feasible.

192b CFD Simulations to Address Mixing in the Transition Region

Quinn Taylor, Julia Baker, David G. Foster

Mixing is a common practice used in the manufacturing and processing industry and is typically modeled in the laminar or the turbulent regime because the equations governing these regimes are well understood.[1-5] Due to the complexity of mixing in the transition regime, minimal modeling exists despite this type of mixing being a common working regime in many industrial processes.[6] Certain industries, including agrochemicals, pharmaceuticals, cosmetics, food and mining, naturally find themselves operating in and around the transition regime. The limited ability to fully understand these processes hampers accurate predictions of optimized configurations that typically involve complex non-Newtonian fluids.

This project utilizes computational fluid dynamics (CFD) as a tool to enable the modeling of various physical systems to analyze velocity profiles, streamlines, and numerous other fluid and flow parameters. Our approach will discretize the governing equations of fluid flow using finite volume methods that solve the Navier Stokes equations numerically via iterative procedures. Using CFD, we will model and analyze properties of fluid domains, such as velocity profiles, pressure contours, and volume fractions. Inaccuracies of transition mixing models may arise from the inability of CFD to predict large scale instabilities in the flow domain. We will circumvent this limitation by incorporating multiphase transient simulations of the mixing of Carbopol, a high molecular weight, crosslinked polyacrylic acid polymer, to study the transient mixing region with high accuracy.

Work in our lab thus far has confirmed that the laminar viscous model is sufficiently accurate when used with fluids in the transition regime using mid to low Reynolds Numbers.[7] The Reynolds Stress Model and Shear Stress Turbulence Model behave well for mid to high Reynolds Numbers as they better capture turbulence within the flow.[7]

In a physical mixing tank several flow regimes exist simultaneously. At the impeller, turbulent flow is likely to exist. In the bulk and the surface of the tank, the fluid can be in laminar flow. Between the two extremes, there will be a wide range of Reynolds Numbers, all falling into the transitional flow regime. The goal of the work was to use CFD simulations to find the best fit model that most accurately represents mixing in all flow regimes in a tank and verify the percentage mixed by extracting the torque data from the simulation to calculate the power number. The power number is a dimensionless number that relates resistance forces to inertial forces and can ultimately be used to calculate the blend time to ensure the tank simulations are fully mixed with 95 percent homogeneity.[2]

Our work includes laboratory mixing videos that were used to compare to the CFD simulation results. In this case, we work to model the volume fraction distribution of dye that is released in the tank while being mixed at various speeds. By following the dye distribution, flow patterns become easier to visualize as well as analyze. Changes of impeller velocity and fluid viscosity enable changes in the Reynolds number. After using a series of equations to find shear rate, shear stress, and viscosity, the necessary impeller speeds were determined to achieve five regimes: laminar, laminar-transition, transition, transition-turbulent, and turbulent regimes. Other than the impeller speed, identical conditions for these regimes were used across the laminar viscous, Reynolds Stress, and k-kl-omega models to select the most accurate scenario.

In the laminar viscous model, the transition between each of the five regimes was clear and smooth while the other two models either yielded no mixing patterns or images that were indistinguishable between regimes. After compiling images of the simulated tanks at various stages of mixing, the Laminar Viscous model was determined to show the best mixing patterns throughout the entire tank.

The result of this work is a more complete understanding of the transition flow regime that is important in the understanding of many industrial mixing processes and the simulation of the various flow regimes that occur simultaneously in various areas of a mixing tank.

References

(1) Oldshue, J. Y.; Herbst, N. R. *A Guide To Fluid Mixing*; Mixing Equipment Corporation, 1990.

(2) Paul, E. L.; Atiemo0Obeng, V.; Kresta, S. M. *Handbook of Industrial Mixing*; Wiley-Interscience, 2004.

(3) Kresta, S. M.; Etchells, A. W.; Dickey, D. S.; Atiemo-Obeng, V. A. *Advances in Industrial Mixing*; John Wiley and Sons: Hoboken, NJ, 2016.

(4) Zhang, Y.; Gao, Z.; Li, Z.; Derksen, J. J. Transitional flow in a Rushton turbine stirred tank. *AIChE Journal* **2017,** *63* (8), 3610.

(5) Machado, M. B.; Bittorf, K. J.; Roussinova, V. T.; Kresta, S. M. Transition from turbulent to transitional flow in the top half of a stirred tank. *Chemical Engineering Science* **2013,** *98*, 218.

(6) Mendoza, F.; Bañales, A. L.; Cid, E.; Xuereb, C.; Poux, M.; Fletcher, D. F.; Aubin, J. Hydrodynamics in a stirred tank in the transitional flow regime. *Chemical Engineering Research and Design* **2018,** *132*, 865.

(7) Baker, H. J.; Yu, W.; Flower, R. D.; Foster, D. G. In *AICHE National Conference* Boston, Mass, 2021.

192f Experimental and Computational Characterization of Mass Transfer in High-Turndown Bioreactors

John Thomas

A systematic investigation into free surface mass transfer and cell growth in high-turndown single-use bioreactors is presented. This report, which combines experimental mass transfer measurements with numerical simulation, deconvolutes the combined effects of headspace mixing and the free surface convective mass transfer on cell growth. Under optimized conditions, mass transfer across the interface alone may be sufficient to satisfy the oxygen demands of the cell culture. Within the context of high-turndown bioreactors, this finding provides a counterpoint to traditional sparge-based bioreactor operational philosophy. Multiple monoclonal antibody-producing cell lines grown using this high turndown approach showed similar viable cell densities to those cells expanded using a traditional cell bag rocker.

192d Simulating Complex Mixing Configurations within a Single Numerical Framework

Lyes Kahouadji, Juan Valdes, Fuyue Liang, Seungwon Shin, Jalel Chergui, Damir Juric, Omar Matar

Mixing flow for industrial applications often exhibits a multitude of physical phenomena such as aeration, emulsion formation, particle clustering, as well as cavern formation in yield-stress fluids. Despite the fact that enormous progress in terms of computational methods has been made for each of these phenomena, it is also rare that a single numerical framework is capable of capturing them all in detail. This talk will highlight our innovative numerical algorithms able to accurately handle: (i) complex geometrical constructions (impellers, baffled tanks, and static mixers); (ii) cavern formation when a stirred vessel agitates viscoplastic materials [Russell et al., 2019]; (iii) aeration when dealing with challenging rapid air-liquid mixing systems [Kahouadji et al., 2022]; (iv) emulsions in liquid-liquid mixing systems [Liang et al., 2022, Vald es et al., 2023]; and (v) particle transport in multiphase mixing systems.

Within our framework, any numerical configuration which necessitates the construction of complex geometrical shapes uses a technique that circumvents the need for time-consuming construction, meshing, and re-meshing of these geometries. Instead, our method proceeds in a modular manner that enables the user to build any geometrical shape from primitive geometrical objects (spheres, cylinders, planes, torii, and ellipsoids) using static distance functions that take into account the interaction of these objects with the flow for both single and multiphase flows.

As an example of mixing with non-Newtonian fluids, “cavern” formation occurs when a viscoplastic material is agitated by a central impeller within a vessel. Outside of this rotating region, the material is stagnant because the stresses being imparted on it are not sufficiently large to exceed its yield stress. This is important as most industrial applications need to minimise stagnant zones in order to achieve efficient, complete mixing. In this situation, our numerical framework has shown remarkable capability of reproducing experimental measurements and excellent agreement with scaling laws involving a combination of dimensionless parameters that incorporate rheological, inertial, and geometrical effects.

It is well-known that the difficulty accompanying the computation of multiphase flows is often due to the high density ratio between the two fluids, which leads to difficulty in convergence of the pressure solver. Our high-fidelity hybrid Level-set/Front-Tracking approach has demonstrated the capability of reaching density ratios of the order of $O(10^5)$ [Shin et al., 2017, 2018]. The implementation of a direct forcing method (for a rotating impeller), coupled with the Large Eddy Simulation Smagorinsky-Lilly turbulence model, has enabled detailed prediction of the challenging aeration phenomena along with an accurate estimation of air hold up and bubble size distribution.

For smaller density ratios, which characterise liquid-liquid mixing, drop size analysis in terms of the temporal evolution of the number of drops, and their size distribution is easily predicted by our framework as a function of dimensionless parameters that reflect the significance of inertia, viscosity, capillarity, geometry, and the presence of surfactants. The addition of solid objects necessitates the introduction of no-slip conditions for these immersed solids; these conditions are enforced taking into account rigid body motion due to their natural translation and rotation.

In conclusion, a combination of innovative numerical algorithms in a single numerical framework, perfectly adapted for the simulation of a wide range of mixing configurations is presented. We demonstrate the capacity of our framework in handling a multitude of multiphase fluid mechanics phenomena in both slow and fast flow regimes.

Acknowledgment: O.K.M. acknowledges funding from PETRONAS and the Royal Academy of Engineering for a Research Chair in Multiphase Fluid Dynamics, and the EPSRC MEMPHIS (EP/K003976/1) and PREMIERE (EP/T000414/1) Programme Grants. We acknowledge HPC facilities provided by the Research Computing Service (RCS) of Imperial College London for the computing time.

D.J. and J.C. acknowledge support through computing time at the Institut du Developpement et des Ressources en Informatique Scientifique (IDRIS) of the Centre National de la Recherche Scientifique (CNRS), coordinated by GENCI (Grand Equipement National de Calcul Intensif) Grant 2022 A0122B06721.

References

L. Kahouadji, L. Liang, J. P. Valdes, S. Shin, J. Chergui, D. Juric, R. V. Craster, and Matar. O. K. The transition to aeration in turbulent two-phase mixing in stirred vessels. Flow, 2(E30), 2022.

L. Liang, L. Kahouadji, J. P. Valdes, S. Shin, J. Chergui, D. Juric, and Matar. O. K. Numerical study of oil-water emulsion formation in stirred vessels: effect of impeller speed. Flow, 2(E34), 2022.

Andrew W. Russell, Lyes Kahouadji, Karan Mirpuri, Andrew Quarmby, Patrick M. Piccione, Omar K. Matar, Paul F. Luckham, and Christos N. Markides. Mixing viscoplastic fluids in stirred vessels over multiple scales: A combined experimental and cfd approach. Chem. Eng. Sci., 208:115129, 2019.

Seungwon Shin, Jalel Chergui, and Damir Juric. A solver for massively parallel direct numerical simulation of three-dimensional multiphase flows. Journal of Mechanical Science and Technology, 31(4), 4 2017.

Seungwon Shin, Jalel Chergui, Damir Juric, Lyes Kahouadji, Omar K. Matar, and Richard V. Craster. A hybrid interface tracking-level set technique for multiphase flow with soluble surfactant. Journal of Computational Physics, 359:409–435, 4 2018.

J. P. Valdes, L. Kahouadji, L. Liang, S. Shin, J. Chergui, D. Juric, and O. K. Matar. Direct numerical simulations of liquid-liquid dispersions in a smx mixer under different inlet conditions. Chem. Eng. J., 46:142248, 2023

192e Towards a General Framework for Development of CFD-Based Reduced-Order Compartment Models (Invited Talk)

Guanyang Liu, Benjamin Wilhite

While computational fluid dynamics (CFD)-based compartment models have become popular as a cost-effective alternative to full CFD modeling of complex mixing systems, developing these models requires substantial trial-and-error efforts. This study introduces a generalized zoning framework that facilitates the streamlined development of CFD-based compartment models, offering detailed characterization of the reduced-order representation of flow physics. By using a stirred tank as an illustrative example, a reduced-order model for species transport and heat transfer with turbulent flow is derived. Subsequently, the generalized zoning framework is presented to illustrate how a simplified CFD simulation can be used to construct a reduced-order compartment model. To evaluate the CFD-based compartment model, a test case involving the mixing of two miscible thermal fluids is conducted. The results demonstrate that the proposed zoning framework exhibits an accurate representation of the CFD simulation of hydrodynamics and demonstrates capabilities of capturing species and heat transfer in turbulent flow systems with complex geometric configurations.

313a Characterizing the Performance of a Rocking Bioreactor for Efficient Mixing and Oxygen Transfer Using a Novel Computational Framework

Minki Kim, Radu Cimpeanu, Daniel M. Harris

Bioreactors are essential for the production of various biological products using cell culturing, such as vaccines, bioplastics, pharmaceuticals, and cultivated meat. In the context of producing cultivated meat, rocking or wave-mixed bioreactors have emerged as a promising innovation due to their disposable nature, low operating expenses, and scalability. However, despite these advantages, the performance of the rocking bioreactor is not well characterized in view of its short history in the market, and the wide range of geometrical and operating parameters.

In the present study we quantitatively evaluate mixing, oxygen transfer, and shear stress within a rectangular rocking bioreactor with different operating conditions based on an implementation in the Basilisk open-source platform. We use the second-order finite volume Navier-Stokes solver and a volume-of-fluid interface reconstruction scheme to accurately resolve the highly nonlinear fluid motion. The non-inertial reference frame fixed into the rocking geometry is used for the formulation of the Navier-Stokes equation, which leads to the addition of non-inertial acceleration terms in the momentum balance equation. By solving the advection-diffusion equation for soluble tracers and oxygen transfer in the multi-fluid system, we examine the degree of mixing and oxygen transfer coefficient (kLa).

For the simulation the rectangular geometry is defined with an aspect ratio of 3.5 that represents a Cell-tainer 2L bag. The bioreactor is operated with a sinusoidal periodic motion, and the operating conditions range from 3 degrees and 10 rpm to 7 degrees and 37.5 rpm for the rocking degree and frequency, respectively. To avoid violent rotating motion at the start and mimic practical motions in experiments, we weakly accelerate the bioreactor with small amplitudes until it reaches the regular amplitude. The tracers and oxygens are released after the bioreactor reaches the periodic state.

Specifically, we highlight two critical hydrodynamic phenomena that affect the performance of the bioreactor. Firstly, we investigate the transitional regime from laminar to turbulent flow, generally unpreferable for cultivated meat production. Moreover, we identify certain operating conditions within the laminar flow regime, which significantly enhance mixing and oxygen transfer. Our findings are expected to provide valuable guidelines for designing optimized bioreactors for the next generation of cultivated meat industry pipelines.

313b Computational Determination of Blend Time in an Industrial Mixing Vessel for Viral Inactivation of an Active Biopharmaceutical Ingredient

Chadakarn Sirasitthichoke, Ishaan Shandil, Piero Armenante

Many biopharmaceutical processes incorporate a viral inactivation (VI) step to eliminate the cytotoxicity associated with possible residual contamination of viral agents. This process is often conducted by adding a viral-specific agent capable of inactivating the virus to the solution containing the Active Biopharmaceutical Ingredient (ABI), such as a protein. Sufficient contact time must be allowed for the blending and inactivation process to take place. In this work a computational approach based on the use of a CFD model utilizing the Lattice Boltzmann Method (LBM) was adopted to quantify the blend time needed to homogenize the viral inactivation agent in an actual full-scale commercial stirred vessel. Accordingly, Triton X-100, a non-ionic detergent solution used for this purpose, was computationally added to the protein solution in the 21.5-m^3, full-scale vessel equipped with an HE-3 impeller and the time for the detergent to become homogenized to the desired uniformity level (95%) was calculated (Figure 1). Simulations were conducted using different addition modalities, such as batch vs. continuous, and different operating conditions, such as agitation speed. Direct experimental validation of the model predictions for blend time was not possible because of manufacturing restrictions associated with the full-scale production in a pharmaceutical plant subject to GMP regulations. However, the results obtained in this study were used to support the successful full-scale production at the biomanufacturing where, as a safety factor, the actual blend time exceeded the blend time predicted numerically here.

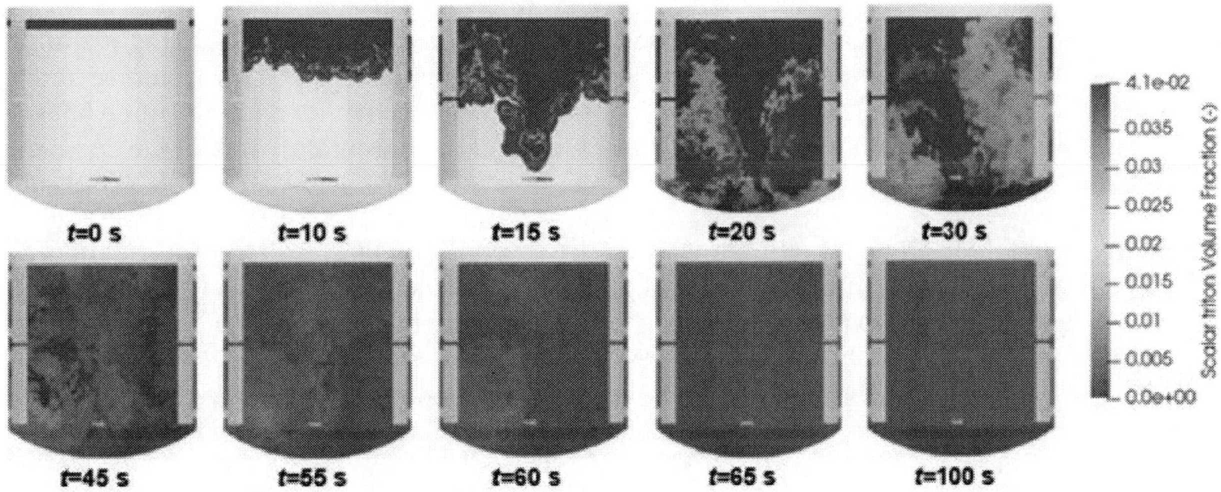

Figure 1. Time evolution of Triton X-100 concentration across the fluid for a single, rapid surface addition of Triton X-100 to the 21.5-m^3 vessel for N=140 rpm.

313c Real-Time Airflow Model for Biopharmaceutical Manufacturing Facility Design

Ahsan Munir, Vikram Pai, Brian To, Yun Tang, John Thomas

In biopharmaceutical manufacturing settings, ensuring the safety and purity of the final drug product is of utmost importance to patient safety. One critical aspect of manufacturing is environmental quality control by maintaining proper Heating, Ventilation, and Air Conditioning (HVAC) requirements in the facility as per the guidelines set by International Society of Pharmaceutical Engineers (ISPE), ISO 14644, and other US and EU guidelines together with establishing a common basis for compliance with current good manufacturing practices (cGMP).

Historically, venting and airflow has been characterized experimentally using smoke studies during clean room and facility qualification to identify any potential sources of contamination. Real time environmental monitoring tools are often deployed to monitor contaminations and assess specific threats to the manufacturing environment which can have a potential impact to product/ patient safety. These tools are effective but resource intensive to implement and expensive to fix if issues are found during system qualification. Additionally, the approach can still have uncertainty associated with airflow patterns and particulate levels in the facility design if traditional design tools are used.

In this work we use transient computational fluid dynamics (CFD) technology based on Lattice-Boltzmann (LB) method to accurately simulate the release and movement of particles in real time, allowing for a more efficient and cost-effective evaluation of airflow patterns. LB is a highly efficient and accurate method for simulating fluid flow and has become increasingly popular in recent years due to its ability to leverage GPU-based computing architectures. By utilizing GPUs, LB simulations can run orders-of-magnitude faster than traditional CFD approaches on CPUs, making it a highly efficient and cost-effective alternative. LB works by discretizing the domain into a lattice and simulating the movement of fluid particles as they collide with each other and the lattice boundaries. By tracking the movement of these particles, LB simulations can accurately model the complex airflow patterns.

By utilizing LB method and integrating it with facility design by using Building Information Modeling (BIM), we have created an airflow digital model of the facility that simulates the real-time air and particulate transport in a manufacturing facility. We are using this model to optimize HVAC design by optimizing the supply and exhaust register locations to minimize steady state particulate levels. Room velocities, temperature gradients, pressure gradients, particulate levels and particulate recovery times are predicted to ensure compliance with room classification criteria during dynamic (operating) conditions.

Overall, the transient CFD models developed in this work will provide an accurate science-based understanding of airflow requirements during the early stages of HVAC and facility design and reduce the overall costs/ timeline for implementation of CapEx projects. Integration of this technology in the critical areas of the facility will be used as risk-assessment tool to provide safe, compliant, and functional systems and facilities for biomanufacturing.

313e Mixing Scale-up for Aerobic Fermentations: From Lab to Pilot to Plant

Richard Grenville, Richard Cope

Scale-up is critical to the successful commercialization of many chemical and biochemical processes. Failure to scale-up accurately will delay plant start-up and, consequently, product launch. Time and money must then be spent to rectify the operational plant-scale issues. This presentation discusses principles used to successfully scale between lab and plant scale mechanically agitated fermenters.

Processes are developed in the laboratory at small scale, typically 10's of liters, then scaled-up to a pilot-plant, typically 1000's of liters, and ultimately commercialized in a fermenter with an operating volume of 100000's liters. The key parameters which must be held constant across the scales are the dissolved oxygen concentration, determined by the organisms' requirements, and the oxygen uptake rate, determined by the desired growth and production rates.

A common method for scaling-up the gas flow rate is to maintain the ratio of the volumetric flow to liquid volume as constant. The gas flow rate can be expressed as vessel volumes per minute, or VVMs. If the oxygen consumption rate is the same across scales, this ensures that the composition of the exiting gas stream and the oxygen concentration gradient are the same.

Finally, there is some evidence that there is a minimum scale at which experiments can be conducted to assess mass transfer performance. The reason for this is that at small scales, there is significant surface motion which draws air down and this contributes to the concentration gradient with the air being fed to the fermenter and this results in the observed mass transfer coefficient being higher than would be expected with no surface motion. The rule-of-thumb recommendation is that the minimum scale is in the range 0.5 – 0.6 m.

122a Optimization of Twisted-Tape Element Static Mixers for Industrially Relevant Setups

Kumar Aanjaneya, Smita Agrawal

Static mixer studies in literature have generally looked at idealized systems with a concentric injected flow. These do not account for real-world mixing scenarios where the additive comes in through an injector. Depending on the injector geometry, the injected flow, and pipe Reynolds number, the injected flow may end up off-center. This impacts the static mixer performance as the first few elements are unable to provide efficient mixing. This work is a parametric study of twisted-tape mixers to numerically ascertain their performance for Newtonian flows. It focuses on industrially relevant mixers which operate with some inexactness, e.g., off-center flow of the additive. Effect on mixing of angle of twist and aspect ratio of the individual element along with Reynolds number are studied; and the Reynolds number dependency of mixer geometry for optimal performance is established. This is done without using some oft-used approximations/assumptions in previous literature like zero-thickness mixer walls, 2d analysis, periodic boundary conditions etc. Two non-dimensional quantities, i.e., Coefficient of Variation (CoV) and Z-factor, are used to assess and quantify the operation of the mixers. For idealized systems, increasing the number of elements by reducing the aspect ratio and/or the twist angle is seen to improve mixing for all Reynolds numbers tested (1-100). However, for a real-world mixer constrained by its total length/pressure drop, mixing is proven to be an optimization problem between increasing the number of flow-splits by introducing a greater number of elements and mitigating the anomalous flow (uneven split) caused due to it. For $Re \leq 10$, reduction of twist angle from 180° to 140° improves mixing for only aspect ratios greater than approximately 0.9. For Reynolds number of 100, the 140° twist angle outperforms the 180° twist angle for all aspect ratios studied. In summary, there is no combination of twist angle and aspect ratio that is a universally optimal solution, and these parameters must be decided subjectively based on multiple other input parameters for a given system.

122b CFD-Based Evaluation of Mixing Efficiency and Flow Uniformity in an Smx Static Mixer

Athanasios Kritikos, George Tsilomelekis, Fernando Muzzio Sr., Ravendra Singh

Imatinib has shown to effectively suppress SARS-CoV-2 infection, and optimizing its production process can contribute to the availability of this drug in the fight against COVID-19.[1] Imatinib production requires the efficient mixing of aniline Imatinib and acid chloride Imatinib with a static mixer. It is essential to optimize the mixer's design and operating conditions, as the product's quality and yield are dependent on its mixing efficiency.[2] Simulations of Computational Fluid Dynamics (CFD) can provide invaluable information regarding the flow dynamics, mixing performance, and optimization of the mixer's design and operating conditions.[3, 4] This research seeks to develop a mathematical model for analyzing the mixing performance of an SMX static mixer and optimizing its design and operating conditions for the production of Imatinib.

CFD was utilized to examine radial mixing and generate RTD profiles in a front-injected tubular mixer with 12 mixing components. A mesh independence test was used to validate the developed model, ensuring the accuracy of the simulation results. In a Sulzer SMX static mixer, the flow dynamics under various conditions and the mixing efficiency along the length of the mixer were analyzed. The velocity, pressure drop, and RTD profiles were examined, and definitive screening analysis was conducted to determine the effect of different solvent properties and operational conditions on mixing efficiency.

The developed mathematical model and CFD simulations contributed to the optimization of similar systems by providing valuable insights into the operation of static mixers. This study's findings can guide future research on SMX mixers and enhance our understanding of the mixing process, thereby enhancing the efficiency and performance of these systems in a variety of applications. The findings can be used to improve the mixing efficiency and product yield of the static mixer used in the production of Imatinib by optimizing its design and operating conditions.

Acknowledgement:

This work is supported by the US Food and Drug Administration (FDA) under contract number 75F40121C00106.

1. Strobelt, R.; Adler, J.; Paran, N.; Yahalom-Ronen, Y.; Melamed, S.; Politi, B.; Shulman, Z.; Schmiedel, D.; Shaul, Y., Imatinib inhibits SARS-CoV-2 infection by an off-target-mechanism. Sci Rep 2022, 12 (1), 5758.
2. Nyande, B. W.; Mathew Thomas, K.; Lakerveld, R., CFD analysis of a kenics static mixer with a low pressure drop under laminar flow conditions. Industrial & Engineering Chemistry Research 2021, 60 (14), 5264-5277.
3. Azizi, F.; Abou-Hweij, W.; Lebaz, N.; Sheibat-Othman, N., A numerical evaluation of flows through an SMX-Plus mixer. Chemical Engineering Research and Design 2022, 178, 382-394.
4. Albertazzi, J.; Florit, F.; Businі, V.; Rota, R., Mixing Efficiency and Residence Time Distributions of a Side-Injection Tubular Reactor Equipped with Static Mixers. Industrial & Engineering Chemistry Research 2021, 60 (29), 10595-10602.

122c *Hydrodynamics and Heat Transfer in Stirred Tanks with Liquid Solid Particulate Flows Using the Lattice Boltzmann Method*

Navraj Hanspal, John Thomas

Multiphase liquid solid particulate flows involving heat transfer in stirred tanks are widely used in fermentation, emulsification, crystallization, homogeneous / heterogeneous reactions, pharmaceutical and agrochemical processes. Heat transfer involved in such operations is extremely critical as it can affect the reaction thermal runaway behavior or the desired particulate based product quality. In this work, we apply a Lattice Boltzmann based Euler-Lagrange CFD approach coupled with Discrete Phase (DPM) and Discrete Element (DEM) Models to simulate the transient temperature distribution of the solid particles, jacketed vessel, and the fluid within the Stirred Tank Reactors through accounting for the inter-particle, fluid-particle and vessel wall-fluid heat transfer. A series of modeling case studies including a Stirred Tank and a Conical Screw Mixer are presented and discussed. The key benefits of using the Euler-Lagrange approach over a conventional Euler-Euler framework includes reduced simulation run-times, and generality of the applied physics, without a need for extensive parameter tuning.

122f A Design of Experiment (DOE)-Based Approach for Bioreactor Scale up and Cell Metabolic Reactions

Sravan Nallamothu, Sivasubramani Krishnaswamy, Hossam Metwally, Sophie Collin, Clementine Shao, Marc Horner

Bioreactors are at the heart of biopharmaceutical manufacturing, and they need to be accurately characterized for optimal scaleup. The transition from lab scale to pilot or production scale with the goal of maintaining the same product yield and homogeneity brings lot of challenges. To evaluate the issues related to scaleup in cell culture bioreactors, one must consider mixing time, oxygen mass transfer rates and shear-stress levels among other factors. As the reactor scale increases, designers would like to ensure consistency in oxygen mass transfer rate as well as exposure of cells to shear because of their influence on the physiology of the cells.

Optimization-based scale up

In this study, computational fluid dynamics (CFD) simulations are used to predict the mass transfer coefficient (kLa) and shear rates for a lab-scale bioreactor. A response-surface (RS) based optimization algorithm is used to optimize the operation of the bioreactor to maximize the oxygen mass transfer rate while maintaining cell exposure to shear under a specified threshold.

The optimized lab-scale bioreactor is then scaled up to pilot-scale. Similarly, a DOE is set up to scan the design space of the pilot-scale bioreactor. Similarly, a DOE will be used to create a response-surface (RS) which is used to optimize the bioreactor operation, maximize oxygen mass-transfer, and minimize shear exposure.

Cell Metabolism System Model

For both scales, the CFD DOE simulations used to construct the response surfaces will also be used to create reduced-order models (ROM's). The ROM's will take the bioreactor operating conditions as inputs (e.g. rpm and sparge rate) and providing the mass transfer coefficient (kLa) as an output.

Moreover, a 0D transient biological cell growth model was developed that predicts the cell growth, product formation, and gas consumption using the mass transfer coefficient (kLa) predicted from the CFD-based ROM. The 0D transient cell growth model is then combined with the CFD-based ROM's in a system model that predicts cell growth, product formation, and gas consumption vs. Time as a function of bioreactor conditions.

The modeling approach proposed herein is an alternative to scale-up laws which may not always be valid (for example when different impeller types are used at different scales). Additionally, the same DOE used for optimization is used to construct ROM's that are used for system modelling which enables the co-simulation of the flow physics and cell biology.

122e *In-Silico* Investigation of Temperature-Controlled Bioreactors

Martin Fruhwirth, Michael Wagner, Philipp Eibl, Christian Witz, Johannes G. Khinast

The design and optimization of bioreactors are critical for efficient and cost-effective production processes. Numerical simulations play a crucial role in predicting the behavior of these complex systems. In contrast to experimental investigations of stirred tanks, computational analysis offers high-resolution spatial data, rapid analysis of different internal geometries, and applicability to different reactor scales once validated. Improving the accuracy and efficiency of current models and schemes will accelerate the optimization process of bioreactors and reduce the need for laboratory tests.

The dynamic behavior of large temperature-controlled tanks can be of particular interest since the integrity of the production process can be highly temperature sensitive. An accurate prediction of the thermal state within the reactor is a key insight in order to avoid local temperature hotspots [1]. In our project, a state-of-the-art implementation of the lattice Boltzmann method is used to simulate the fluid flow and the heat transfer in a jacketed stirred vessels for biopharmaceutical fermentation. Both the lattice Boltzmann framework as well as the transport scheme for thermal energy can utilize advances in graphical processing units (GPUs) and therefore ensure fast execution times due to the high degree of parallelization.

These models then simulate the separate flows of the liquid within the stirred vessel and the heat transfer medium within the jacket. The heat transfer through the jacket is modelled with thermal wall functions and therefore accounts for the resistance on both sides. A large eddy scheme is applied for turbulence. The results are validated with experimental data of lab-scale reactors at different flow conditions and show a promising agreement for large scale design processes, both qualitatively and quantitatively. The results are also compared with multiple Nusselt-correlations.

In addition, the simulation of both fluid flows enables a separate analysis of the flow conditions within the vessel and the jacket like the inlet velocity or inlet temperature during the heating and cooling procedure. Different flow conditions in the outer heating jacket are analyzed and how they affenct the overall time required for heating and cooling of the media inside the stirred vessel.

[1] Cummings, Cody M. et al. "Transient thermal modeling of bioprocess equipment", International Journal of Heat and Mass Transfer 184, 2022, 122064

This talk will be in the online proceedings, with all links live.

I will give a similar talk to the Process Development Division
as part of their quarterly webinar series on March 14, 2024

"Process Always Comes First"
when successfully applying mixing technology

A personal reflection on 40 years
in industrial mixing and process development

Aaron Sarafinas
Sarafinas Process & Mixing Consulting LLC

2023 AIChE Annual Meeting – Orlando, FL – Paper 90B

Abstract

- Understanding how mixing affects a process is critical to bridge the gap between the lab and the plant. We must avoid surprises on scale-up to achieve commercial success.
- I have been very fortunate in my career to have the opportunity to work on many challenging chemical and physical processes. Many of those required us to balance multiple competing rate processes and sometimes conflicting process responses. Almost always, some of the multiple competing rate processes and process responses would be affected by mixing. Since various mixing parameters scale up differently, understanding which mixing parameters controlled the various process responses and how those would scale up was usually critical to the commercial success of the project.
- In this Special Lecture, I will give a personal reflection on my 40 years in industrial mixing and process development. I will share some examples from my career where mixing was critical to commercial success, give my perspective on the critical connection between process and applied mixing technology, relate my experience with networking as an effective way of managing (and disseminating) technology within a company, and talk about ways to efficiently characterize how mixing affects chemical and physical processes (the Bourne Protocol and more).
- Our understanding of mixing technology has grown greatly over the last 40 years. In the examples, I will highlight how we approached certain problems over the years and why some of these "old" approaches should not be forgotten despite recent technological advances. I will close by giving my thoughts on current trends in process development and mixing, along with my opinions on future needs for the mixing and process development communities.

I start my mixing course with this pre-test...

- In scale-up,
  what are the three scales of interest?

Do NOT shout out your answer.

I'll come back to this later...

- There is an obvious answer, but there is also a more informed answer.

In less than the next hour…

- Introduction
- Suspension polymerization got me addicted to mixing
- Networking as an effective way of gathering, managing, and disseminating technology within a company
- 3 quick examples of how process always comes first
- Ways to efficiently characterize how mixing affects competitive chemical and physical processes: the Bourne Protocol and more
  - Fixing a precipitation leads to a fixation with the Bourne Protocol
- Current trends in process development and mixing
- Final thoughts

Very light on equations
Very heavy on process

My introduction – over 35 years with Rohm and Haas / Dow

1982	1991	2000	April 1, 2009		May 2018
ROHM AND HAAS			DOW®		Sarafinas Process & Mixing Consulting LLC
IER Research	Ag Process Research	Engineering Technical Center	Core R&D / Engineering & Process Sciences	Advanced Polymers Process Technology	
Res Process Engineer	Sr Process Engineer / Group Leader	Technical Fellow		Global SME	Principal & Mixing Steward
Polymerization and Polymer Processing	Reaction systems and many unit operations	Internal consultant to all business areas	PD&A FM&M	Polymerization / polymer proc	
Process Development & Scale-up / Scale-down					
Mixing Technology					

I will be using this graphic to show how the agenda items aligned with year, business area, job title, and technology areas. At all times, I was working on Process Development & Scale-up/scale-down with a focus on mixing technology.

My introduction

Always been tall for my age

- Really a process development person, specializing in mixing
  - "Good mixing" is impossible without the proper process context.
  - Use the right tools in the right way at the right time to build process understanding and achieve commercial success.
- Process developer who was bitten by the mixing bug...
  - Development, scale-up, commercialization, and troubleshooting for many products and processes for a variety of businesses
    - Starting with scale-up of suspension polymer reactors
    - Many different multiphase contacting and reaction systems
    - Building process understanding and simplifying processes
  - Developed and delivered training on scale-up and mixing technology
  - Technical presentations at various AIChE meetings and contributor to the NAMF handbook <u>Advances in Industrial Mixing</u>

2023 AIChE Annual Meeting – Orlando, FL – Paper 908

6

My introduction, continued…

- Elected to retire from Dow in May 2018
- Started consulting company enabling clients to
  - Achieve faster process development and scaling, and
  - Leverage advantaged mixing technology
- Services:
  - Experienced Problem Solving
  - Faster Process Development
  - Does Mixing Matter?
  - Simpler and Improved Processes
  - Managing Technical Capabilities
  - Delivering Technical Training

"helping clients understand how mixing matters in their process and how to deal with mixing effects to avoid unpleasant surprises on scale-up"

For more info:
http://sarafinasprocess.com

Some thoughts on becoming a consultant

- My reply on AIChE Discussion Central (10-26-2019) included some things that I found helpful in considering independent consulting
  - Dave Dickey webinar (AIChE Academy)
  - Chemical Consultants Network
  - Laura Burford "Laura's Consulting Guide"
  - Learn how to take care of "everything"
  - Follow work processes
    - Establish your "focus"
    - Client needs, proposal, collaboration
    - Finances and business
- This has been fun (for the most part…)

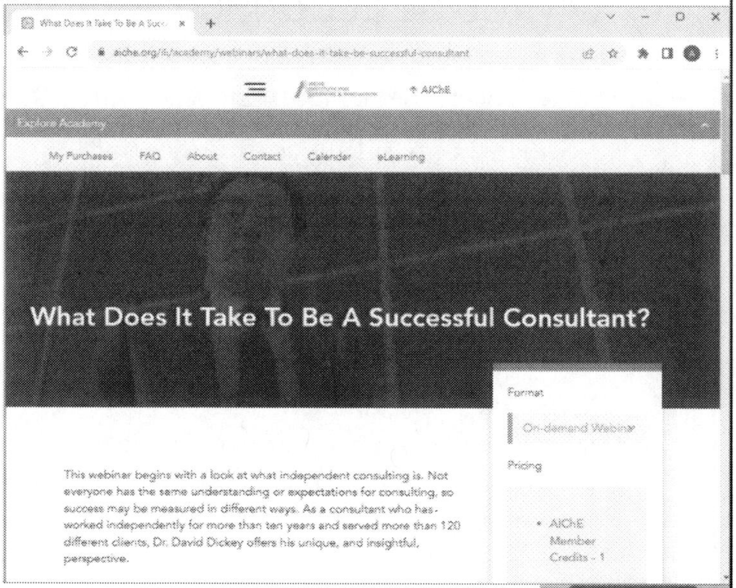

2023 AIChE Annual Meeting – Orlando, FL – Paper 908

Links:
Dave Dickey webinar: https://www.aiche.org/ili/academy/webinars/what-does-it-take-be-successful-consultant

CCN: https://chemconsultants.org/

Laura's Consulting Guide: https://www.laurasconsultingguide.com/

AIChE Discussion Central post: https://engage.aiche.org/communities/community-home/digestviewer/viewthread?MessageKey=dcce6567-53bf-44d7-b50f-07779ced9e68&CommunityKey=de21ae11-bfdc-41b4-851d-84c7293f0d99#bmdcce6567-53bf-44d7-b50f-07779ced9e68

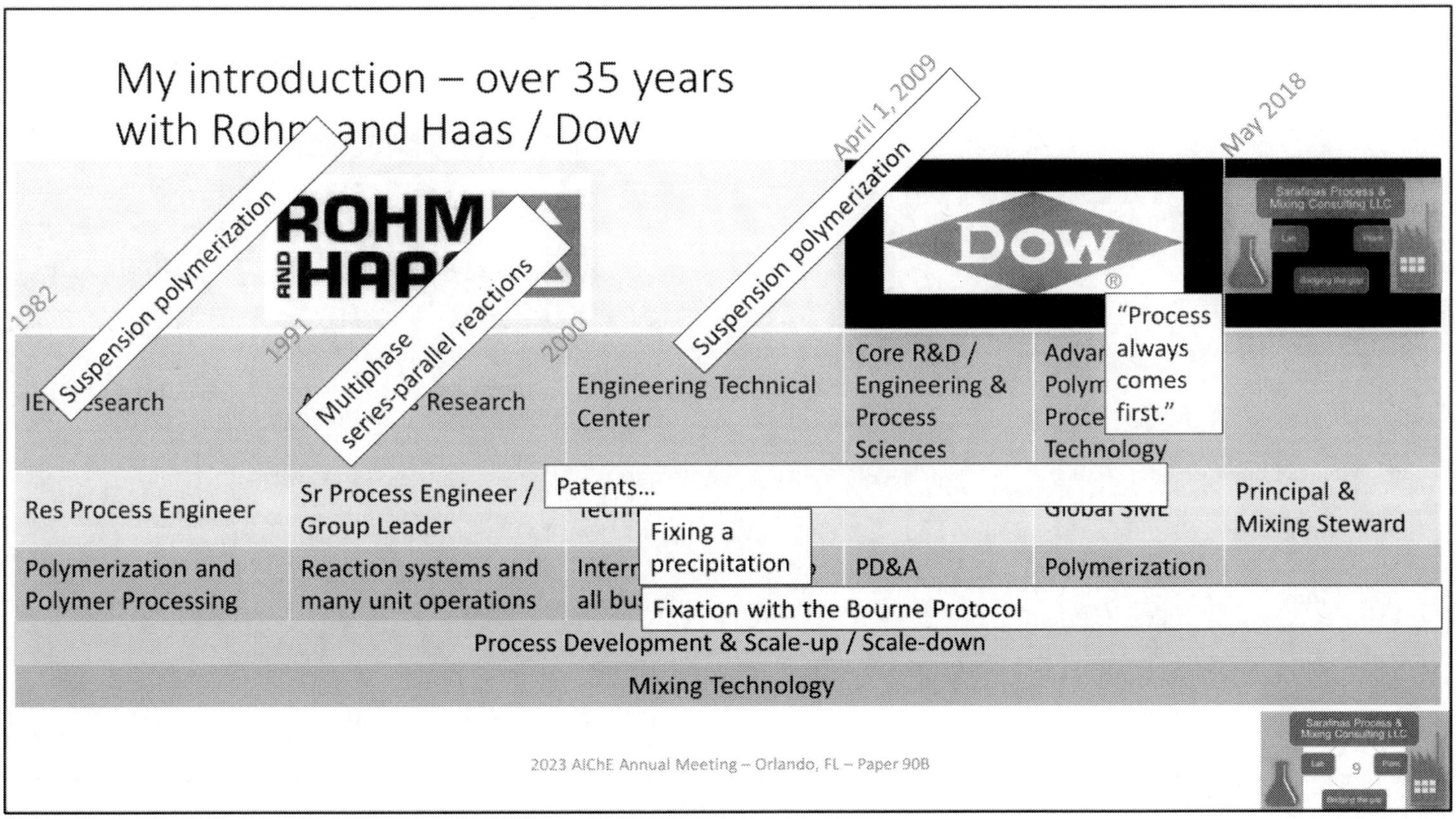

My introduction – over 35 years with Rohm and Haas / Dow
ROHM AND HAAS
DOW
Sarafinas Process & Mixing Consulting LLC
1982
1991
2000
April 1, 2009
May 2018
Suspension polymerization
Multiphase series-parallel reactions
Suspension polymerization
"Process always comes first."
IE Research
Research
Engineering Technical Center
Core R&D / Engineering & Process Sciences
Technology
Res Process Engineer
Sr Process Engineer / Group Leader
Patents…
Fixing a precipitation
Principal & Mixing Steward
Polymerization and Polymer Processing
Reaction systems and many unit operations
PD&A
Polymerization
Fixation with the Bourne Protocol
Process Development & Scale-up / Scale-down
Mixing Technology
2023 AIChE Annual Meeting – Orlando, FL – Paper 90B
9

My introduction – over 35 years with Rohm and Haas / Dow

ROHM AND HAAS | DOW | Saratinas Process & Mixing Consulting LLC

1982	1991	2000	April 1, 2009		May 2018
IER Research	Ag Process Research	Engineering Technical Center	Core R&D / Engineering & Process Sciences	Advan Polym Proce Technology	
Res Process Engineer	Sr Process Engineer / Group Leader	Technical Fellow		Global SME	Principal & Mixing Steward
Polymerization and Polymer Processing	Reaction systems and many unit operations	Internal consultant to all business areas	PD&A FM&M	Polymerization / polymer proc	
Process Development & Scale-up / Scale-down					
Mixing Technology					

"Process always comes first."

2023 AIChE Annual Meeting – Orlando, FL – Paper 90B

10

"Process always comes first"

- Conversation with a trusted colleague as I was setting up my LLC
  - My question, "Should I call my business Sarafinas
    - Mixing and Process or
    - Process and Mixing?"
  - Her answer, "Process *always* comes first."
- That was a truth that I realized throughout my career, so…

Process always comes first.

Why are there droplets on the card? More about that in a moment…

"Mixing" dictates that "process always comes first"

- Definition of "mixing" is based on achieving process results
  - From HIM (2004): We define *mixing* as the reduction of inhomogeneity **in order to achieve a desired process result**. (Bold added for emphasis)
  - From *Applied Mixing Technology for Process Development: Scaling Down to Scale Up*

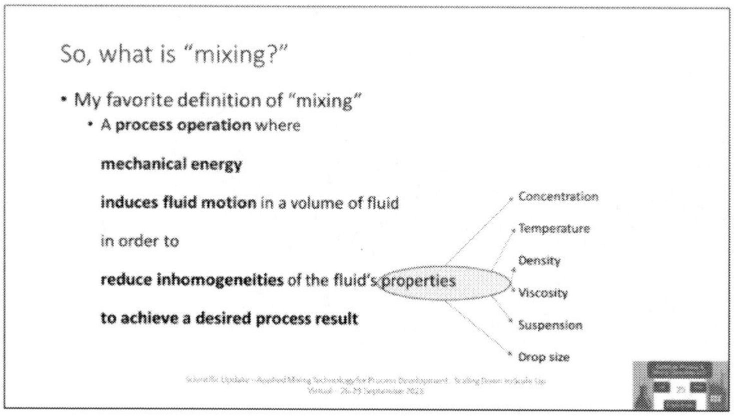

Developing and implementing and troubleshooting and optimizing processes is central to the profession of Chemical Engineering and fluid mixing technology is often critical to process success

My introduction – over 35 years with Rohm and Haas / Dow

1982 · 1991 · 2000 · April 1, 2009 · May 2018

Suspension polymerization

Suspension polymerization

"Process always comes first."

IE Research	Ag Process Research	Engineering Technical Center	Core R&D / Engineering & Process Sciences	Advanced Polymer Process Technology	
Res Process Engineer	Sr Process Engineer / Group Leader	Technical Fellow		Global SME	Principal & Mixing Steward
Polymerization and Polymer Processing	Reaction systems and many unit operations	Internal consultant to all business areas	PD&A FM&M	Polymerization / polymer proc	
Process Development & Scale-up / Scale-down					
Mixing Technology					

2023 AIChE Annual Meeting – Orlando, FL – Paper 908

13

It all started with suspension polymerization…

- What is suspension polymerization?
  - Preparation of aqueous and organic phases (around equal volumes)
  - Liquid-liquid dispersion of initiated monomer in aqueous
  - Avoid coalescence and agglomeration by using "suspending agents"
  - Heat to polymerization temperature
  - Monomer polymerizes exothermically
  - Separate solid polymer from aqueous phase for subsequent processing
    - Classification (desired PSD between 300 and 800 microns, or narrower)
    - Post-polymerization reactions

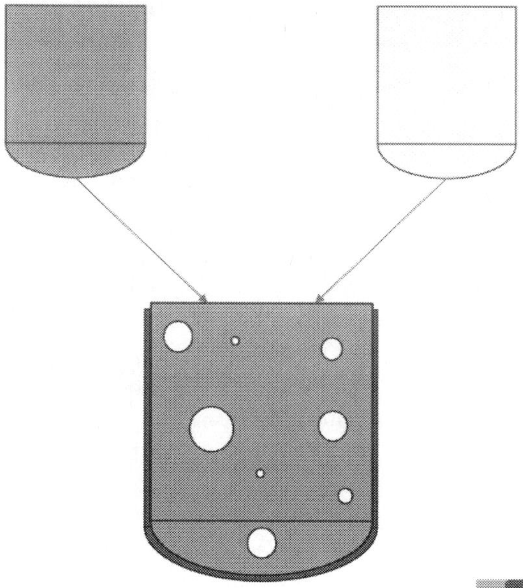

2023 AIChE Annual Meeting – Orlando, FL – Paper 908

14

In a 2,000 gal (8m3) reactor, if all the droplets were 500 microns, then there would be about 20-million drops in the reactor.
With no coalescence, think of the 2000 gal reactor as being filled with 20-million batch microreactors.

It all started with suspension polymerization…

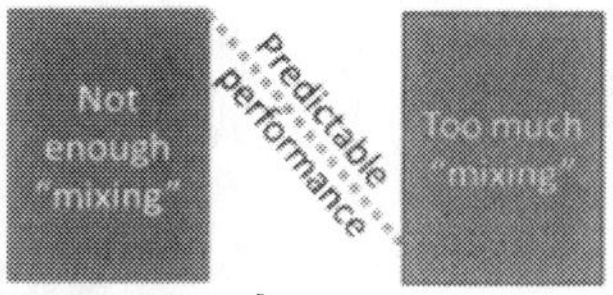

- What is suspension polymerization?
  - Preparation of aqueous and organic phases (around equal volumes)
  - Liquid-liquid dispersion of initiated monomer in aqueous
  - Avoid coalescence and agglomeration by using "suspending agents"
  - Heat to polymerization temperature
  - Monomer polymerizes exothermically
  - Separate solid polymer from aqueous phase for subsequent processing
    - Classification (desired PSD between 300 and 800 microns, or narrower)
    - Post-polymerization reactions

- Mixing applications
  - Blending (phase preparation)
  - Liquid-liquid dispersion
    - Minimum dispersion speed
    - Drop size (average and distribution)
  - Gas-liquid (avoid aeration)
  - Heat transfer (heat up & polymerization)
  - Solid-liquid
    - Off-bottom suspension
    - "uniform" composition to next unit operation

Real processes often have multiple, sometimes uncomplimentary, mixing needs

OK, fine, but why do the work?

- Respond to competitor's increase in production scale
    - A RandH manager heard that Dow had increased suspension polymer production scale to 6,000 gallons by listening to a talk by Doug Leng at an AIChE meeting in the early 1980s
    - RandH chose to investigate increasing production scale, too, but needed to consider
        - How the various parts of the suspension polymerization process would scale
        - How we could improve the particle size distributions
- Project plan
    - What is the state of the technology internally and in the literature?
    - Investigations at various scales in pilot plant on most critical mixing needs
        - Role of impeller type and system design and scale and fluid properties on
            - particle size distributions
            - minimum dispersion speed
        - Measurement of heat transfer coefficients during the polymerization
    - Scale up improved designs to existing production scale
    - Install larger scale reactor(s) with the improved design

Not inventing a new impeller

State of the technology in the literature

- Various and sundry correlations for minimum dispersion speed and d_{32}
  - d_{32} based mostly on the paths blazed by Hinze in the 1950's and Shinnar in the 1960's
    - Calabrese et al papers in the mid-1980's, excellent discussion in HIM (2004)
  - Minimum dispersion speed correlations were all over the place...
    - I prefer the Armenante et al papers from the late-80's to early-90's
- Not much in the literature on particle size distribution
- Two rather interesting papers
  - JY Oldshue, DO Mechler and DW Grinnell (1982) "Fluid Mixing Variables in Suspension and Emulsion Polymerization" **Chem Eng Prog,** May 1982, 68-74.
    - Two plots (with no numbers on the axes) comparing the shear rate distributions at the pilot and production scales each at two different D/T's
    - Implied that D/T must increase on scale up to get the same shear rate distribution

D/T increases on scale-up? What about geometric similarity?

- Some use the photocopier for scale-up:

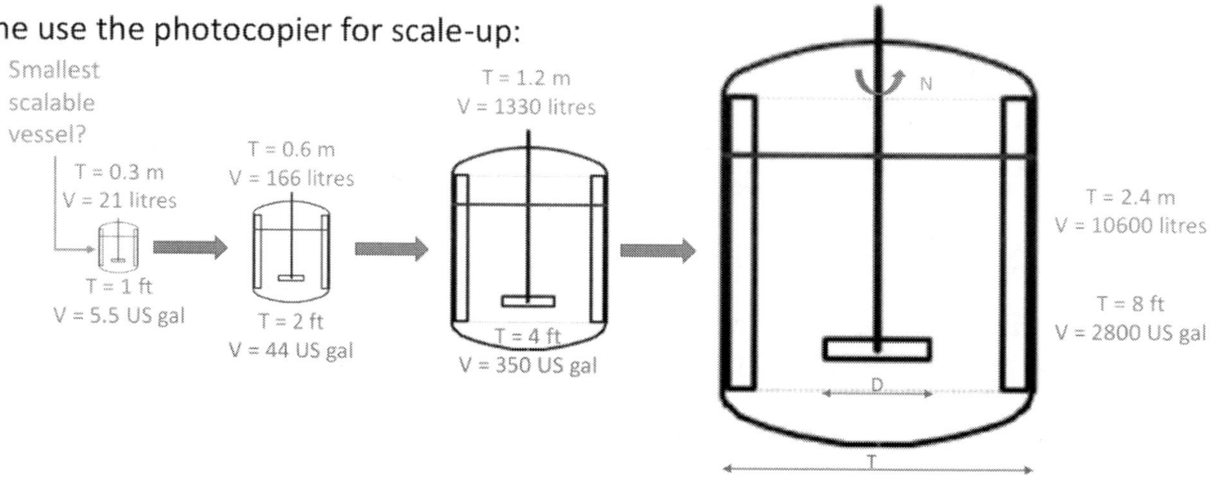

- But Oldshue implied that we might not want to do that

Two quotes on non-geometric similarity - 1

- Oldshue (1983)

It is necessary to sensitize the pilot plant and its parameters by using nongeometric techniques to get the blend time, circulation time, and other phenomena in a range comparable to some degree with what is happening in the plant-size unit. **This may mean making the small-scale mixing tank as poor a mixing device as the plant-size tank, in order to make meaningful studies of variables.** Experience shows that unless this is done, the effect of some variables will be missed completely in the laboratory and pilot plant, and "strange things" will happen on the way to the large-scale tank.

Two quotes on non-geometric similarity - 2

- Leng (1991)
  If processes require both constant power per unit volume and shear to be maintained, conditions of geometric similarity need to be relaxed. **In mini-plants, non-similar geometry is used to simulate mixing in a large scale reactor.** For example, when blend time is critical, mini-plant studies can be done in a reactor having a smaller impeller diameter / tank diameter ratio.

- My conclusions
  - Embrace non-geometric similarity
  - You don't need a scale model, you need a model that scales

State of the technology in the literature

- Various and sundry correlations for minimum dispersion speed and d_{32}
  - d_{32} based mostly on the paths blazed by Hinze in the 1950's and Shinnar in the 1960's
    - Calabrese et al papers in the mid-1980's, excellent discussion in HIM (2004)
  - Minimum dispersion speed correlations were all over the place…
    - I like the Armenante et al papers from the late-80's to early-90's
- Not much in the literature on particle size distribution
- Two rather interesting papers
  - JY Oldshue, DO Mechler and DW Grinnell (1982) "Fluid Mixing Variables in Suspension and Emulsion Polymerization" **Chem Eng Prog,** May 1982, 68-74.
    - Two plots (with no numbers on the axes) comparing the shear rate distributions at the pilot and production scales each at two different D/T's
    - Implied that D/T must increase on scale up to get the same shear rate distribution
  - D.E. Leng & G.J. Quarderer (1982) "DROP DISPERSION IN SUSPENSION POLYMERIZATION," **Chemical Engineering Communications**, 14:3-6, 177-201.
    - Yes, it's from Dow, but they really don't use that loop impeller, do they?
    - Different drop size model because of the loop impeller
    - Excellent discussion of critical suspending agent concentration

The challenge of industrial process research

- Within the project constraints, get a "good enough" answer that will "work"
  - The goal is not elegant research,
    nor models,
    nor published papers,
    nor new agitators to sell
  - "The goal is commercial success" –Kamlesh Bhatia (2002)
  - "Goal Deliberation" or some other sensible scope setting work process
    to define "good enough" (Sarafinas, 2003)

I need to understand how mixing
matters in my process and
how to deal with mixing effects to
avoid unpleasant surprises on scale-up

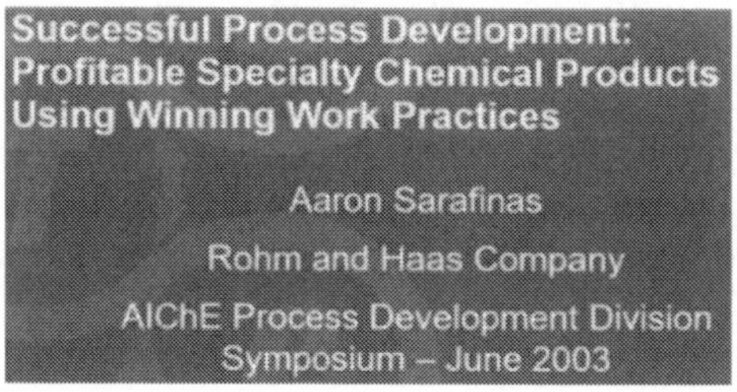

Discovering the solution AND making it work

- Project team saw two potential solutions using existing impellers
- Scaled the less novel design to the existing production scale, results were good, not great
- Studied the more novel design in more detail at the pilot scales
- Scaled the more novel design to the existing production scale with superior results
  - Confirmed inertial subrange model for our systems
  - Narrower particle size distribution

$$\frac{d_{32}}{D} = C_1(1 + b\phi)\left(\frac{N^2\, D^3\, \rho_c}{\sigma}\right)^{-3/5}$$

  - Established proprietary correlations for scaling minimum dispersion speed and surface aeration
    - Appropriate for this style design and these fluids
    - Applicability for other systems and designs?

Another characteristic of industrial research: PROPRIETARY correlations and designs

- Installed the larger scale production vessel with superior results
  - Validated improved particle size distribution, minimum dispersion speed, and surface aeration
- Learned so much, and very satisfying to see it all work…
  I was hooked on mixing

2023 AIChE Annual Meeting – Orlando, FL – Paper 908

23

How would we do this today? What gaps are there still?

- Some thoughts about CFD
  - "It's not ready for that yet." –Midey Chang-Mateu in the mid-1980s on my suspension polymerization work as a potential application for the new PHOENICS CFD tool at RandH
  - For an hour of my thoughts on CFD, check out my discussion with John Thomas in the M-Star podcast "Will CFD Ever Replace Experiment?"
  - Would CFD indicate an improved drop size distribution for the optimal design?
  - How adequately are the fluid properties and interface behavior captured?
- There is definitely a place for "model-guided experimentation"
- There is always a need for doing experiments with the real fluids
- There are alternative drop formation technologies that can make even narrower drop size distributions… the agitator task becomes "don't mess up the DSD"

Links
M-Star podcast: https://mstarcfd.com/will-cfd-replace-experiment-podcast-6/

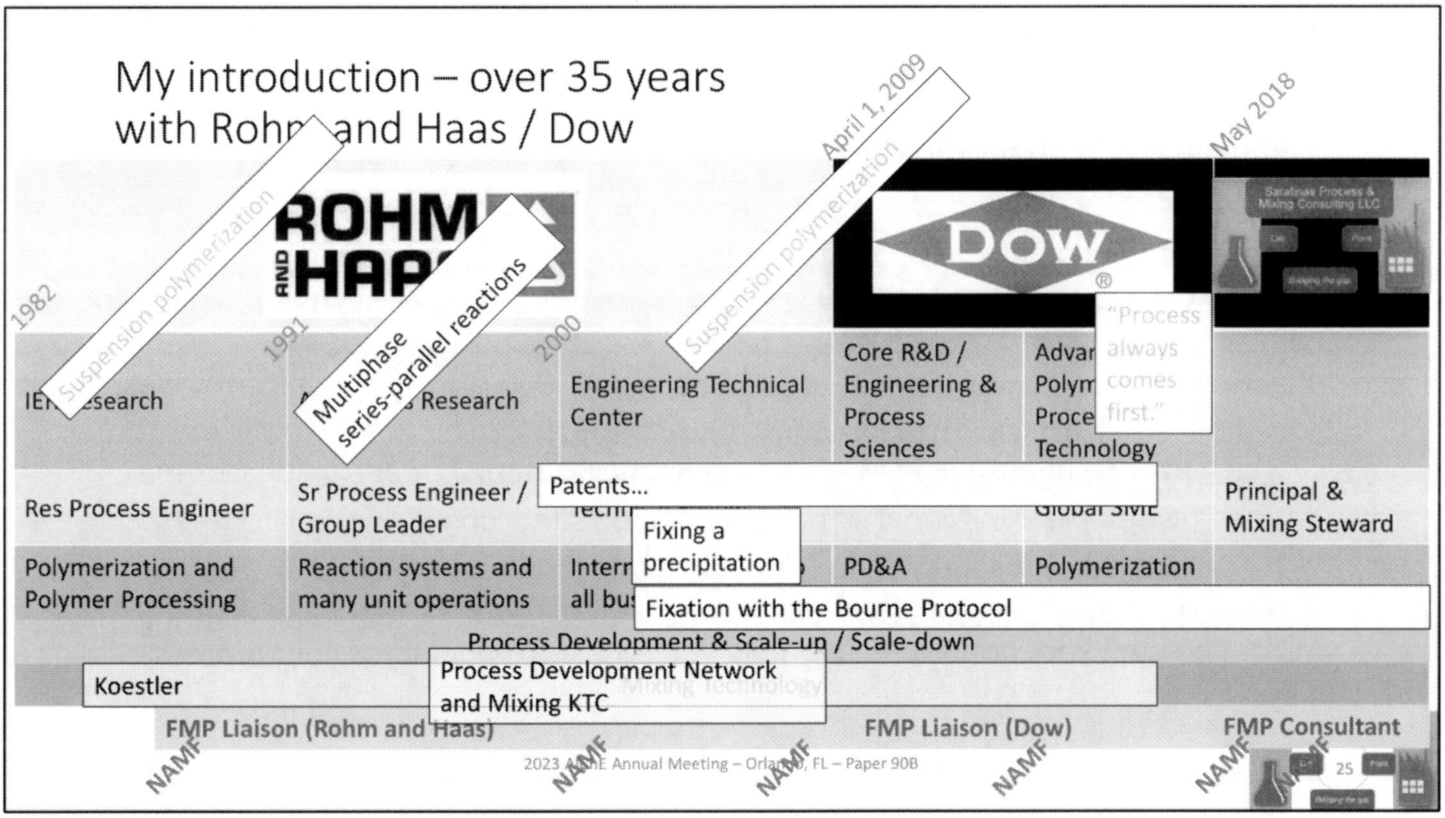

Networking topics: Don Koestler mentoring, FMP involvement, NAMF conference attendance, and Process Development Network and Mixing KTC

Don Koestler – "Understand your process…"

- Emphasized "systems thinking" and never was a "cookbook" engineer
- My mentor for decades, made me "draw the envelope" big enough
- One of the founding members of NAMF
- The inspiration and driving force for the growth of our mixing capabilities at RandH
- Responsible for the creation of our Process Development Network and our "virtual" management of technology
- Got me into the facilitating of brainstorm/peer reviews, which led to a multitude of "Generate & Reduce Options" opportunities for me
- Unsurpassed passion for networking and systems thinking
- "Life is a batch" presentation to PDD: Success = (Process) * (Skills) * (Attitude)
- Don's three keys to success in inventing, developing, scaling, and implementing: "Understand your process, understand your process, understand your process"

Mixing Networking: Koestler and FMP and NAMF

- Don Koestler got me to attend FMP SCMs and NAMF conferences starting in 1987
    - FMP SCMs and NAMF conferences very energizing experiences
- RandH Technical Community Organization (TCO) focused on technical networking
    - Technical Discussion Groups
    - Annual Symposia and Tech Fairs
- Mixing Discussion Group started from urging by Koestler
- Quadrennial Rohm and Haas Mixing Conference was a TCO Symposium
    - Started in 1984, continued to 2000

FMP = Fluid Mixing Processes Consortium

FMP BY FRAMATOME BHR

- Framatome BHR's Fluid Mixing Proceses (FMP) Consortium https://fmp.framatomebhr.com/
- Industrially-driven mixing research and development project
- Funded by member companies since 1983
- Market-leading research on fluid mixing processes
  - Producing design guidelines and scale-up methods
  - Directly applicable to industry
  - Incorporated into FMP Excel Design Guide
  - Available to member companies only
    - For at least 5 years
    - Public release of any items are usually subject to member approval
- Member companies review research programs in Steering Committee Meetings and vote to prioritize FMP's research

"FMP continues to generate **cutting-edge industrially-relevant mixing information** for their member companies that is **years ahead of what is available in the public domain**.

"As always, the steering committee meetings were excellent opportunities for members to learn about FMP's latest research and give input into FMP's proposed research that can best meet member needs. The dialogue and the insights were thought provoking."
--Sarafinas LinkedIn post 4/6/22

2023 AIChE Annual Meeting – Orlando, FL – Paper 90B

Sarafinas Process & Mixing Consulting LLC 28

Members: 22 end users, 5 equipment manufacturers
Link: https://fmp.framatomebhr.com/

My history with FMP

- Liaison from Rohm and Haas / Dow to FMP since the late 1980's
  - Role of the liaison
    - Manage the relationship between the company and FMP
    - Attend Steering Committee Meetings to hear research results and give input to programs
    - Conduit for information from FMP to the company
    - Manage the company votes to help prioritize the FMP research programs
  - FMP materials were the key external mixing technology knowledge used at RandH/Dow
- Became an "FMP Consultant" when I started consulting in 2018
  - Give input to FMP on research programs
  - Participate in Steering Committee Meetings
  - Have access to all FMP materials
  - No voting privileges in the project prioritization (but I'm very vocal with my opinions)
  - Not part of FMP's "two weeks of consultancy each year" for member companies
- Attended just about every US steering committee meeting since late 1980s (since they were held in Philadelphia)

My opinion on FMP

- From my presentation at the FMP 40th anniversary celebration (October 2023)
  - No one else in the process industries delivers what FMP delivers
    - Content: Industrially-relevant mixing research disseminated to members
      - Areas – wide breadth:
        blending, solid-liquid, liquid-liquid, gas-liquid, reactive mixing, static mixers, jet mixers
      - Scales – wide breadth: T=1ft (and smaller) to T=9ft
    - Connection: Unique networking opportunity for mixing community
      - Listening to members for most relevant research areas
      - Dialogue to focus research and interpret results
  - There are many opportunities for FMP to improve the state of mixing knowledge

AIChE Process Development Division and Symposium

- NAMF started in the 1980's, PDD became a division around 2000
- Started presenting at AIChE in 2000 in PDD sessions
- Process Development Symposium
  - Chair of Day 1 at the first PDS in 2003
  - Longtime organizer and facilitator of the "interactive session"

PDD
Process Development Division
An AIChE Community

2023 AIChE Annual Meeting – Orlando, FL – Paper 90B

31

RandH Process Development Network and Mixing KTC

- From Hunt and Sarafinas (2002)
- Community of Practice, not a Center of Excellence
- Designed to be a virtual organization
  - Practitioners, Stewards, Leadership Team
- Made up of talented technical people
  - Research, engineering, and manufacturing
- Creating competitive advantage
  - for all our businesses
  - by our ability to invent, develop, and scale-up chemical processes.

RandH Process Development Network and Mixing KTC

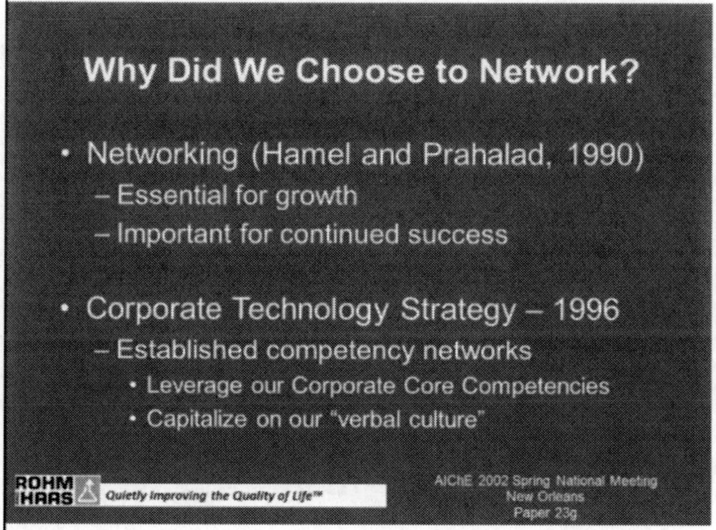

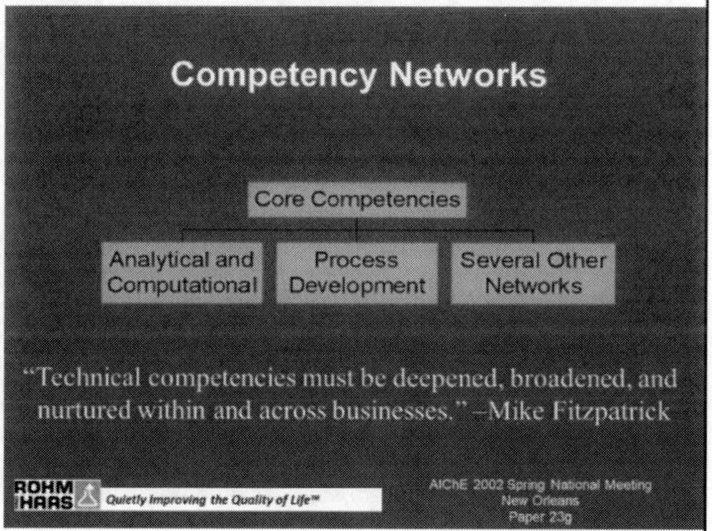

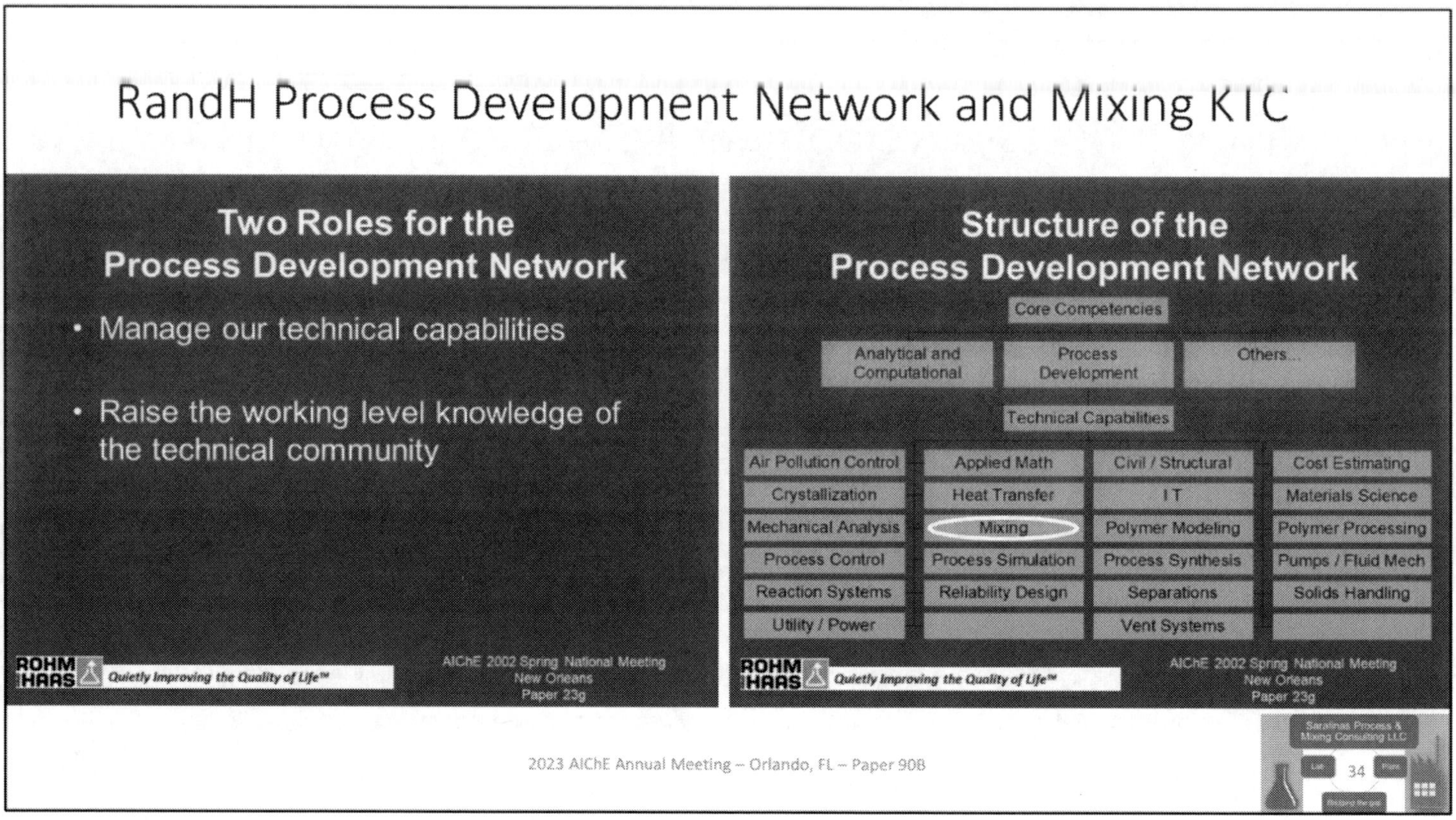

Mixing was the first "Key Technical Capability" group with six Stewards (Aaron Sarafinas, Midey Chang-Mateu, Gloria Miller, Dave Pierce, Pete Edwards, and Don Koestler). Mixing was the most successful of the KTC's.

RandH Process Development Network and Mixing KTC

How Do These Technical Capability Groups Work?

- TC's are managed by Stewards
  - Technical experts within that capability
- Steward Responsibilities
  - Raise working knowledge across company
  - Share and disseminate knowledge and current best practices
  - Be apprised of the best practices
  - Select and support external technical work

ROHM HAAS Quietly Improving the Quality of Life™ AIChE 2002 Spring National Meeting New Orleans Paper 23g

Roles and Duties in the Technical Capability Groups

- Steward Responsibilities, cont'd
  - Maintain technical databases
  - Serve as consultants
  - Define key questions requiring long-term studies
  - Maintain the network
- And, do all of this part time…
  - Stewarding is only 5-25% of worker's job.

ROHM HAAS Quietly Improving the Quality of Life™ AIChE 2002 Spring National Meeting New Orleans Paper 23g

Note that the top responsibility is "raising working knowledge" and that consulting is lower in priority. We wanted to "teach people how to fish, not give them a fish."

RandH Process Development Network and Mixing KTC

And how do Stewards spend that other 75%-95% of their time?

- Most Stewards work in the businesses
  - Business pays for their PDN time
    - Administratively stay in the business
    - Concept: business will benefit from PDN activities of employees from all businesses
    - Varying degrees of success with businesses
- Other Stewards from corporate consulting groups

ROHM HAAS Quietly Improving the Quality of Life™

AIChE 2002 Spring National Meeting
New Orleans
Paper 23g

PDN Leadership

- Leadership Team
  - senior managers in Research, Engineering, and Manufacturing
- Leadership Team Duties
  - continuously improve the operational effectiveness of the PDN
  - support Stewards' needs
  - increase the visibility of the PDN at all levels.

ROHM HAAS Quietly Improving the Quality of Life™

AIChE 2002 Spring National Meeting
New Orleans
Paper 23g

RandH Process Development Network and Mixing KTC

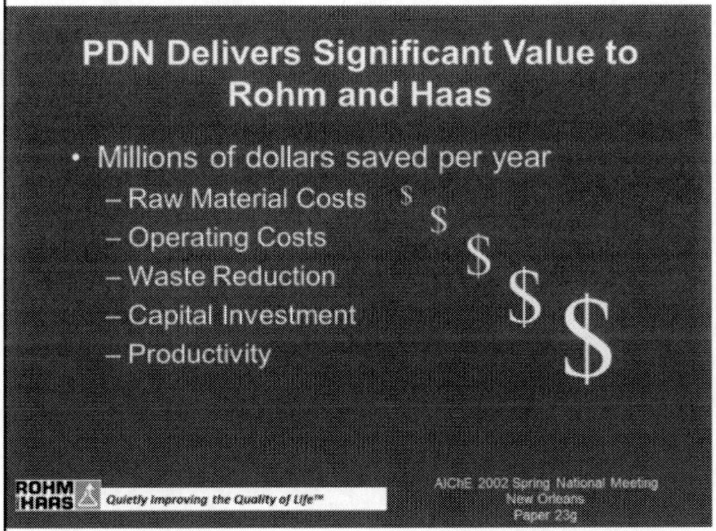

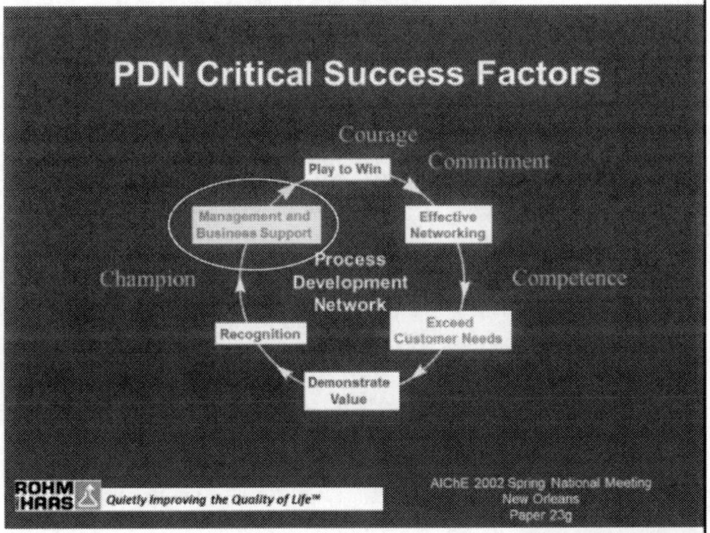

2023 AIChE Annual Meeting – Orlando, FL – Paper 908

The PDN ultimately ceased to exist because of reduced management and business support. The Mixing KTC was the last active group, but the Stewards moved to other jobs, with me going to a Center of Excellence as a process development and mixing expert.

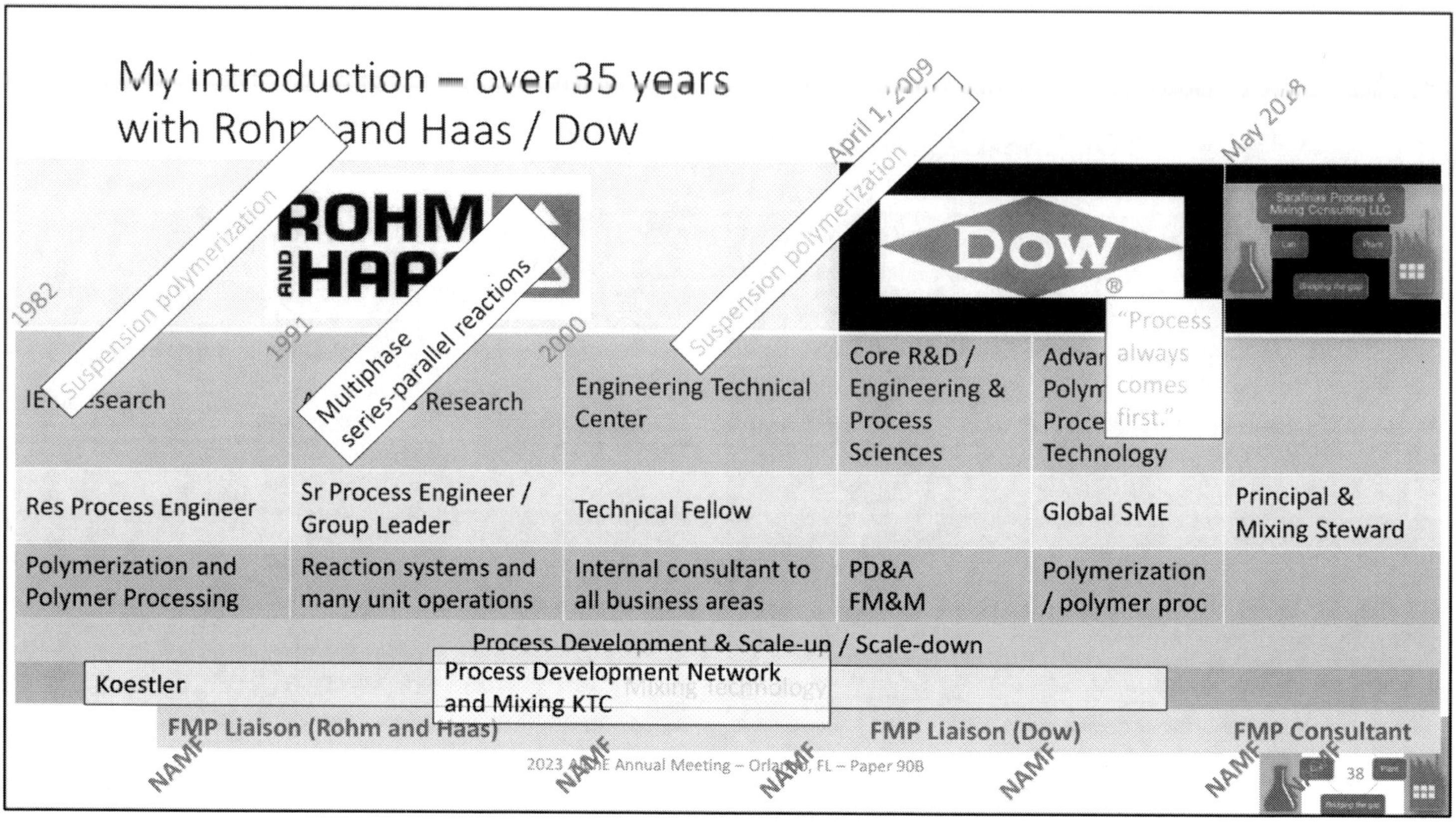

My introduction – over 35 years with Rohm and Haas / Dow
1982
1991
2000
April 1, 2009
May 2018
ROHM AND HAAS
DOW
Suspension polymerization
Multiphase series-parallel reactions
Suspension polymerization
"Process always comes first."
Engineering Technical Center
Core R&D / Engineering & Process Sciences
Technology
Res Process Engineer
Sr Process Engineer / Group Leader
Technical Fellow
Global SME
Principal & Mixing Steward
Polymerization and Polymer Processing
Reaction systems and many unit operations
Internal consultant to all business areas
PD&A FM&M
Polymerization / polymer proc
Process Development & Scale-up / Scale-down
Koestler
Process Development Network and Mixing KTC
FMP Liaison (Rohm and Haas)
FMP Liaison (Dow)
FMP Consultant
NAMF
2023 AIChE Annual Meeting – Orlando, FL – Paper 90B
38

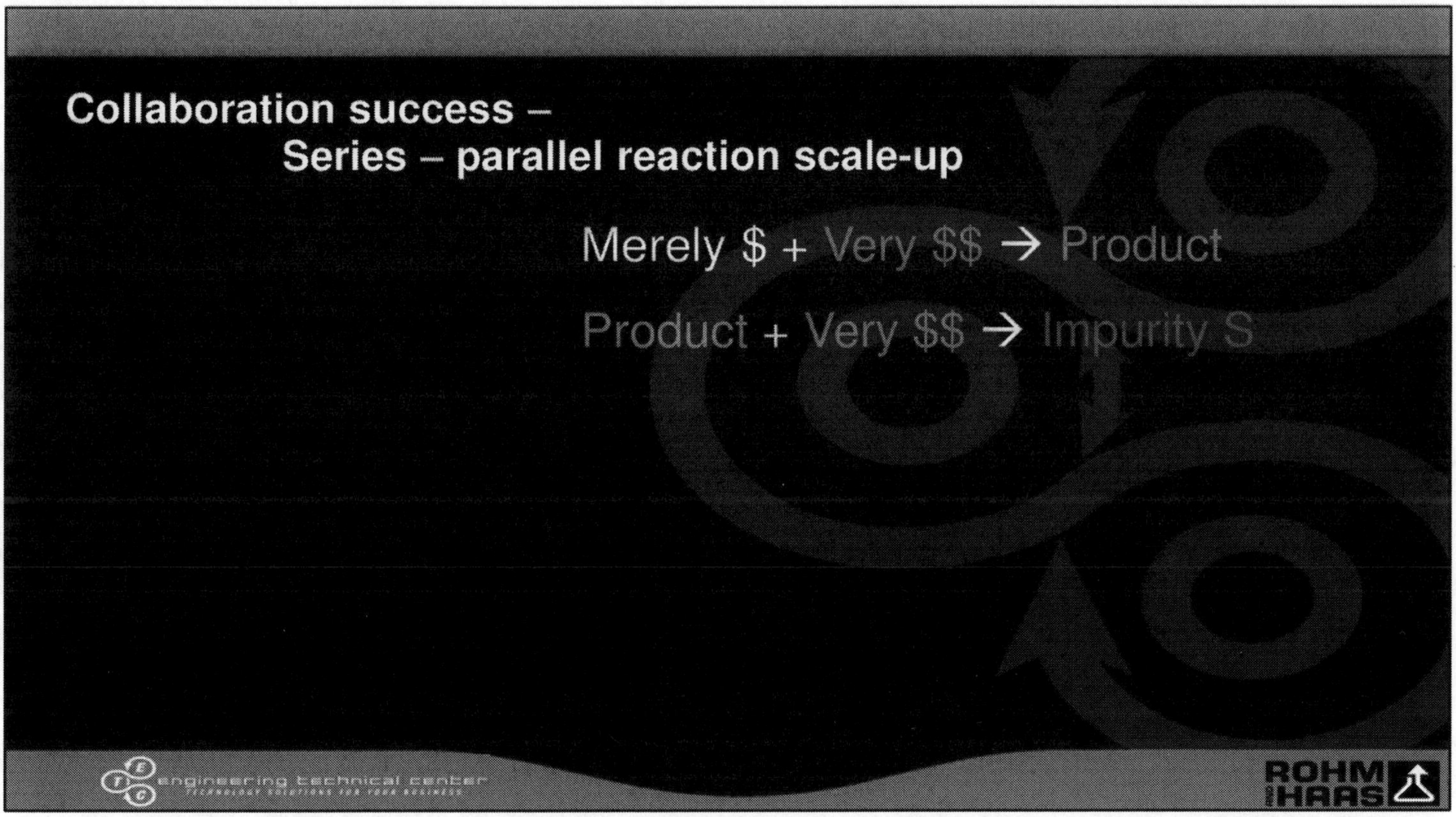

From my PDS presentation, Sarafinas (2003).
In this case, the process involved a reaction of a very expensive raw material with another raw material that was merely expensive. Unfortunately, the "merely expensive" raw material had two functional sites and could consume two moles of the "very expensive" raw material to make an impurity. The engineers recognized that this series-parallel reaction scheme was the subject of micromixing studies in the literature.

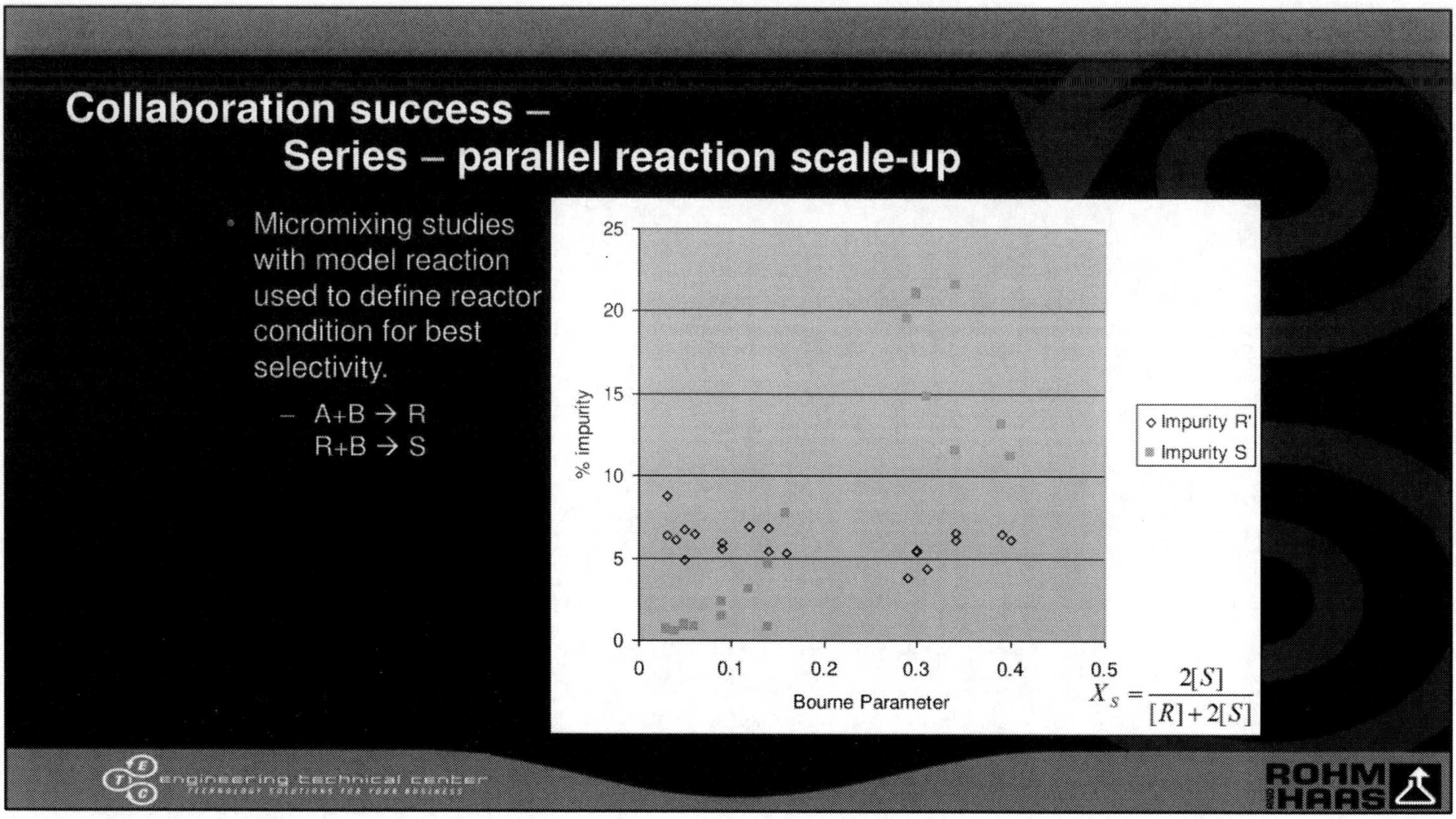

They chose to run some model reactions with a similar series-parallel structure to characterize the micromixing in the lab and see how the micromixing correlated with impurity generation. As shown by the squares in the plot, the "S" impurity correlated well with the "Bourne parameter" (yield loss of B to "bad stuff") that characterizes micromixing, while, as expected, the reverse isomer (reacting on the "wrong" functional group) was independent of mixing.

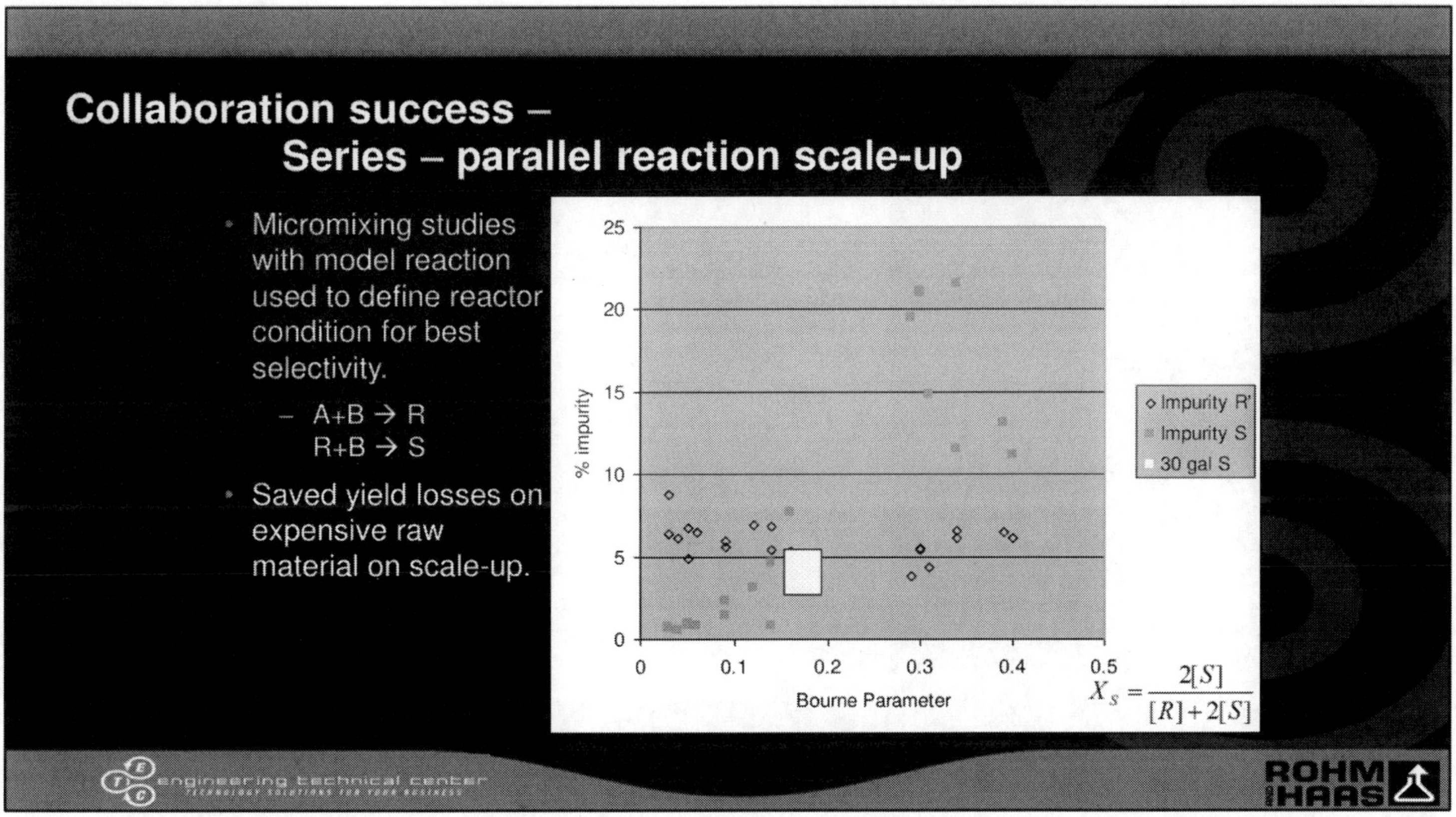

The team also carried out micromixing studies at the pilot scale to determine the optimal reactor conditions and configuration. The S-impurity generation at the 30-gallon scale agreed very well with the micromixing prediction from the lab. The experimental approach taken by the integrated project team made sure that they would not waste the very expensive raw material through an unwise reactor design at the pilot plant.

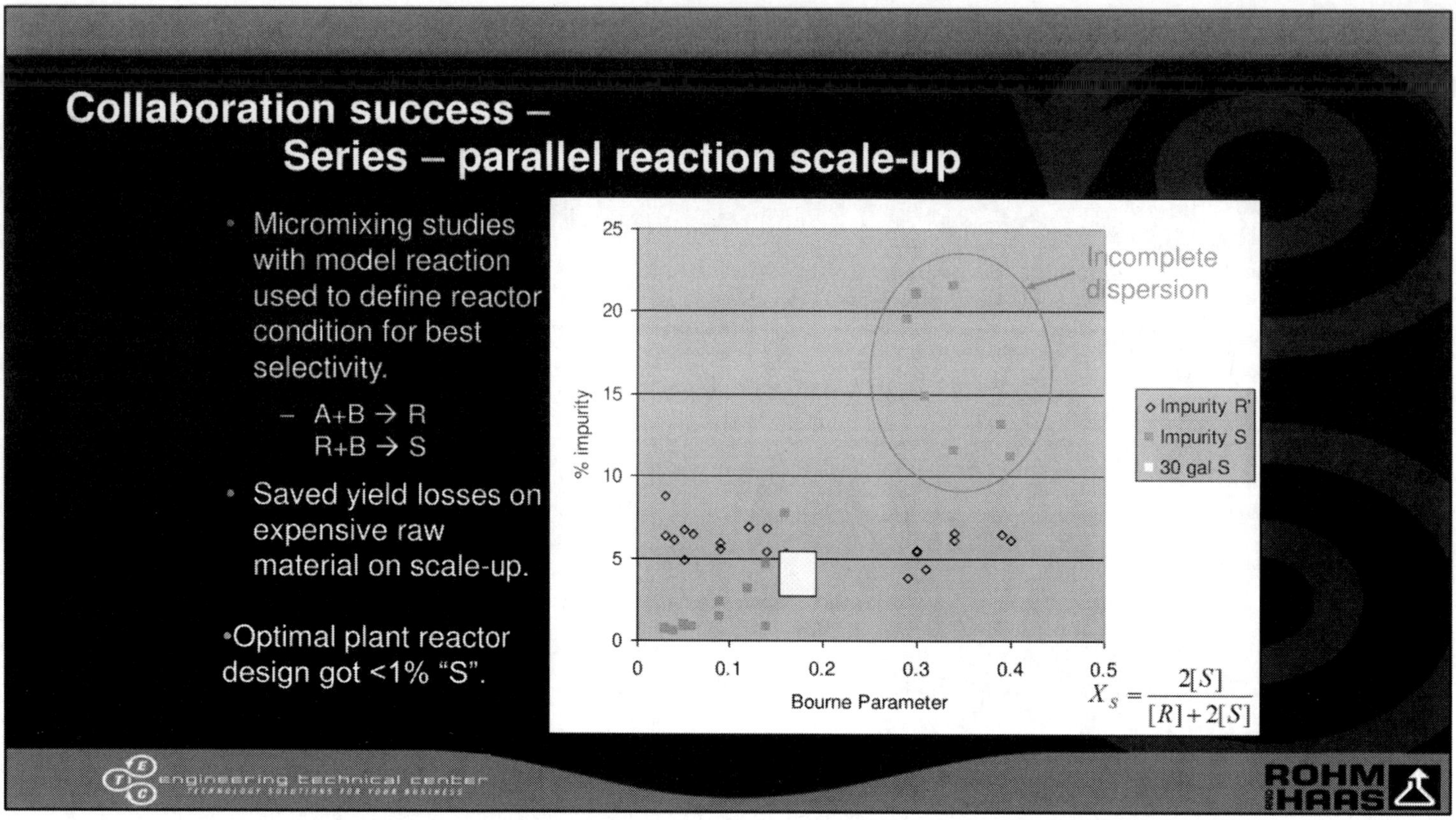

Ultimately, the process was scaled to 300 gallons, where some mass transfer issues were identified, and then to 8 cubic meters. Thanks to the collaborative understanding of micromixing and mass transfer, the production reactor produced less than 1% of the S-impurity.

I'll be discussing an alternative approach to studying this sort of problem with the Bourne Protocol as a way to build better process understanding, faster, which enables higher profitability, sooner.

We would develop that process faster today

- The project team would probably not jumped directly into to doing the experiments with the first Bourne reaction.
- The project team would have started with a Process Scheme to look at the different phases, components, and rate processes at work in the process.
- The project team would have conducted Bourne Protocol experiments and mapped the results into the micromixing-mesomixing operating space.

- Do not be afraid to challenge paradigms

Reaction systems can be complex

- From DynoChem User Group Meeting (Berrier et al 2007):

Using DynoChem® to Reduce Batch Cycle Time in a Gas-Liquid Reactor and Commission a New Unit

John Berrier, David Kolesar, Aaron Sarafinas, and Cheryl Teich
16 May 2007

- Results
  - Model pointed us to move conditions beyond those at existing plants
  - Reduced hold time from 20hr to 12hr
  - Debottlenecked two plants
  - Eliminated catalyst in related process
- Model guided experimentation works…

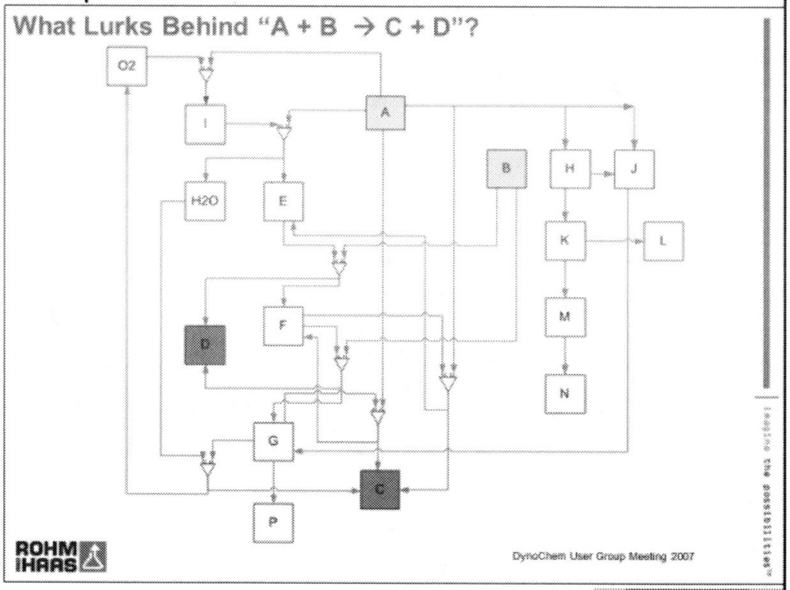

My history with Dynochem® from Scale-up Systems

Scale-up Suite:
“the world’s leading drug substance process development and scale-up software for scientists and engineers working in the pharmaceutical industry.”

(but it’s not just for pharma)

My history with Dynochem® from Scale-up Systems

- Long-time user of Dynochem at Rohm and Haas / Dow
  - I find that Dynochem is a versatile and valuable tool for process development that enables faster process understanding through sensible modeling
  - Applied Dynochem mixing utilities to the Bourne Protocol
- Continued use and collaboration with Dynochem in my consulting
  - Collaborated on project with Dynochem to improve their mixing utilities, which resulted in the Mixing and Heat Transfer Toolbox (MHTT)
    - Gave a three-part webinar series for Dynochem users called "Talking Mixing"
    - Improved the correlations and other methods used in the MHTT based on materials available in the public domain
    - Gave a three-part training webinar series on using the MHTT to do the calculations that I previously discussed in "Talking Mixing"
  - Search for my lectures and training series on Dynochem Resources at https://dcresources.scale-up.com/#q=Sarafinas 6209
- **Congratulations on your retirement, Marion Maniet!**

Link: https://dcresources.scale-up.com/#q=Sarafinas 6209

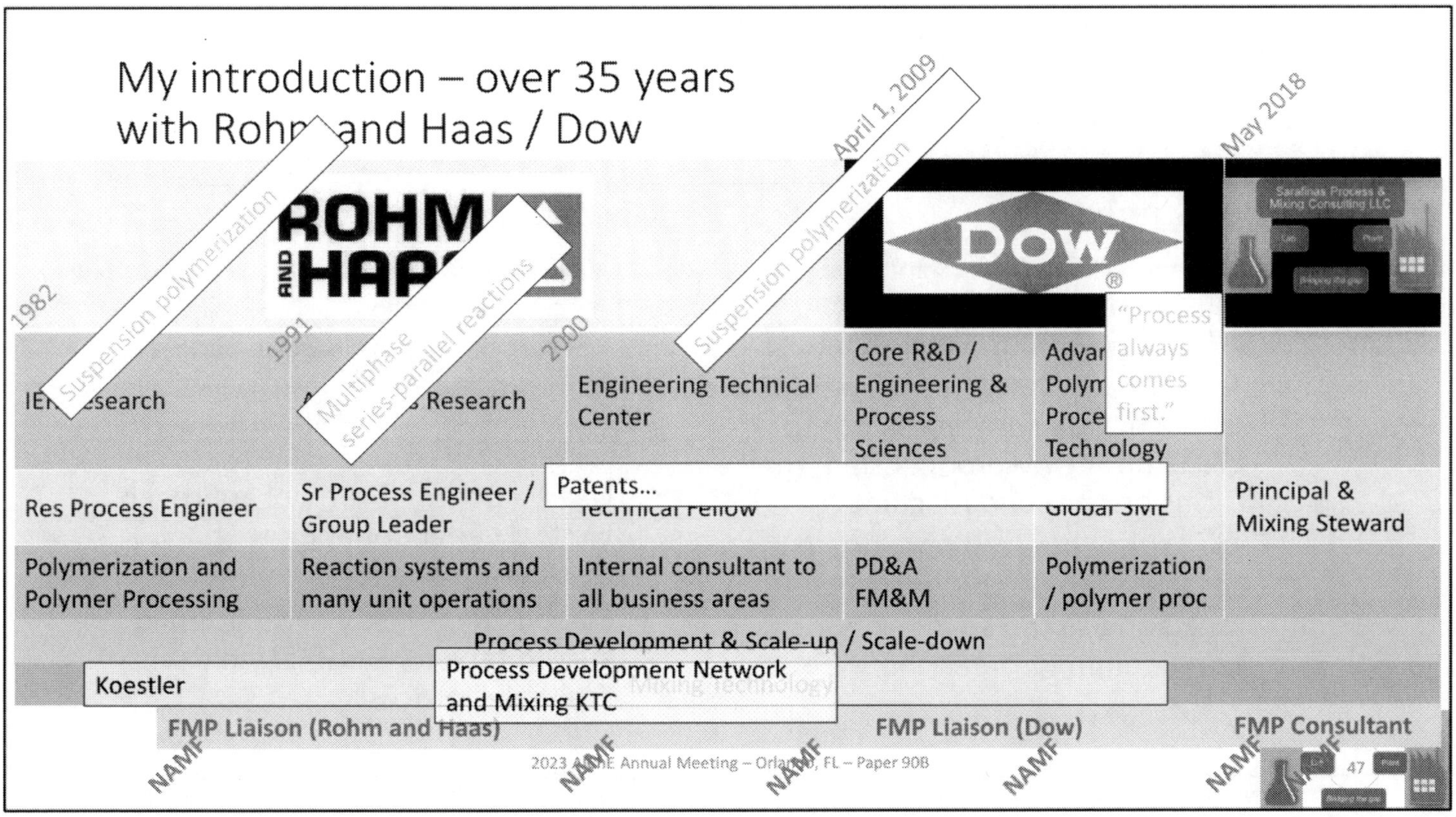

Rohm and Haas changed their philosophy on process patents in the late 1990's. I had 7 process patents for different business areas in the company, with 6 of them related to mixing. I will highlight only one of those.

US-7029641 - Method of making $LiBH_4$

- Process need
  - Chemist: I need to scale up a reaction in a refluxing ball mill, and we want to toll manufacture
  - Textbook method for making $LiBH_4$ uses a ball mill as the reactor to break up the by-product shell that forms on the solid reactant
  - Me (after biting my tongue): let's see if we can do this in a stirred tank and use the solid reactant as the medium to break the shell
- Mixing applications
  - Solid-liquid mixing
    - Solid suspension
    - Particle attrition (particle-particle)
  - Blending & Reactive mixing
- Process result
  - Used a Bourne Protocol approach to study effect of energy dissipation and feeds and reflux
  - Established energy dissipation in stirred tank for optimal yield with "tollable" process

2023 AIChE Annual Meeting – Orlando, FL – Paper 90B

48

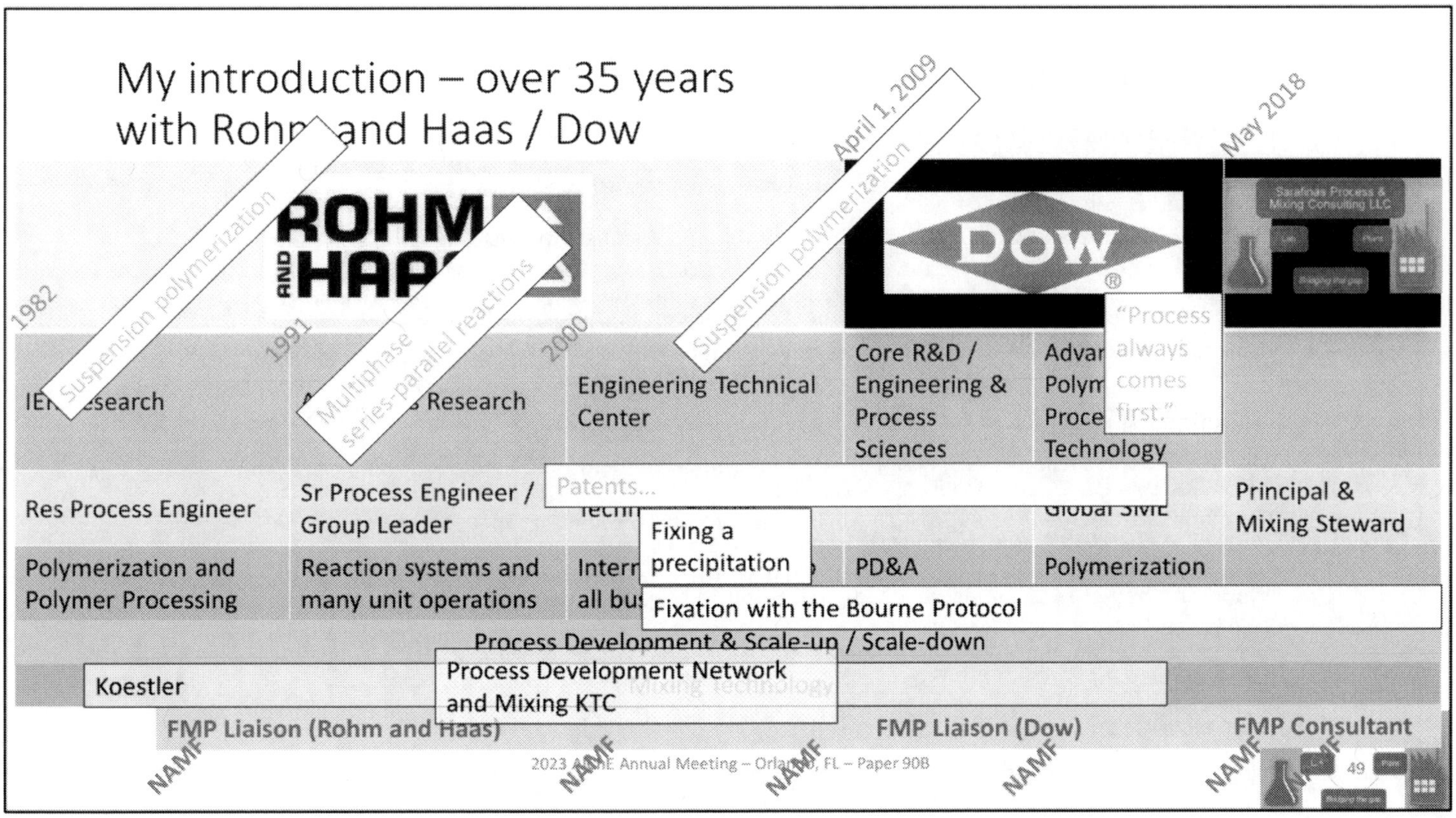
My introduction – over 35 years with Rohm and Haas / Dow
1982
1991
2000
April 1, 2009
May 2018
Suspension polymerization
Multiphase series-parallel reactions
Suspension polymerization
IE Research
Research
Engineering Technical Center
Core R&D / Engineering & Process Sciences
Polym Proce Technology
"Process always comes first."
Res Process Engineer
Sr Process Engineer / Group Leader
Patents…
Principal & Mixing Steward
Fixing a precipitation
Polymerization and Polymer Processing
Reaction systems and many unit operations
PD&A
Polymerization
Fixation with the Bourne Protocol
Process Development & Scale-up / Scale-down
Koestler
Process Development Network and Mixing KTC
FMP Liaison (Rohm and Haas)
FMP Liaison (Dow)
FMP Consultant
NAMF
2023 AIChE Annual Meeting – Orlando, FL – Paper 90B
49

I talk about the Bourne Protocol and the effect of mixing on competitive rate processes for four hours when I teach the Scientific Update course "Applied Mixing Technology for Process Development"

Many Bourne Protocol talks...

- **AIChE 2010 Annual Meeting**, Salt Lake City, UT, November 2010 (2 papers):
  - "Can this process be saved? A search for understanding using the Bourne Protocol and advanced process development tools"
  - "Taking the min to the max: a case study in small scale process development using on-line reaction calorimetry and in-situ particle characterization"
- **Advances in Industrial Mixing: A Companion to the Handbook of Industrial Mixing**, John Wiley and Sons, 2016. (Chapter 13b with Cheryl Teich)
- "A further look at the Bourne Protocol for efficient investigation of mixing sensitivities during process development," **AIChE 2017 Process Development Symposium**, Toronto, Canada, June 2017.
- "Using the Bourne Protocol to test mixing sensitivities in the lab," **DynoChem Guest Webinar**, 8 August 2018. (Available online)
- "Test Process Mixing Sensitivities Using the Bourne Protocol," **Scientific Update Webinar**, 13 November 2018. (Available online)
- "Mixing-sensitive reactions: recognizing them and dealing with them" to Chemical Consultants Network, Sept 2019 (http://www.chemconsultants.org/event-3503651)
- "Efficiently Characterize Process Scalability in the Micromixing-Mesomixing Space Using the Bourne Protocol," **AIChE 2019 Annual Meeting,** Orlando, FL, November 2019.
- Sarafinas, A., "Efficient Process Development in the Micromixing-Mesomixing Space: the Bourne Protocol and Beyond," **AIChE 2020 Annual Meeting**, Virtual, November 2020. (link to video of presentation)
- Sarafinas, A., "Scaling Down to Scale Up: Advances in Using the Bourne Protocol and Optimization in the Micromixing-Mesomixing Space," **AIChE 2021 Annual Meeting**, Virtual, November 2020. (link to video of presentation)
- Sarafinas, A., "Further Adventures in the Micromixing-Mesomixing Space: Next steps with the Bourne Protocol," NAMF MIXING XXVII, Virtual, June 2022. (link to video of presentation)

2023 AIChE Annual Meeting – Orlando, FL – Paper 90B

50

I have been on this Bourne Protocol journey for quite a while, starting at Rohm and Haas long before our first papers at the 2010 AIChE Annual Meeting. I list the public papers and our chapter in Advances in Industrial Mixing here. There are some links to recordings of the presentations or the slides if you want more detailed background, because I'm going to fly through some of the content.

DynoChem Guest Webinar: https://dcresources.scale-up.com/?q=Sarafinas+Bourne%20Protocol#q=Sarafinas+Bourne+Protocol

Scientific Update Webinar: https://www.scientificupdate.com/webinar_events/test-process-mixing-sensitivities-using-the-bourne-protocol/20181113/

CCN presentation: http://www.chemconsultants.org/event-3503651

AIChE 2020 Annual Meeting: https://youtu.be/pCXbyWjwLyM

AIChE 2021 Annual Meeting: https://youtu.be/5S1VF-6Zpn0

NAMF MIXING XXVII: https://youtu.be/jhB_HoXqBrU

Information about Scientific Update "Applied Mixing Technology for Process Development" course:
https://www.scientificupdate.com/training_courses/applied-mixing-technology-for-process-development-scaling-down-to-scale-up-4/20221003/

Think about length scales and time scales

- My reinterpretation of the diagram in Lerou and Ng (1996)
- Wide combination of length and time scales
- What could be important for your process?

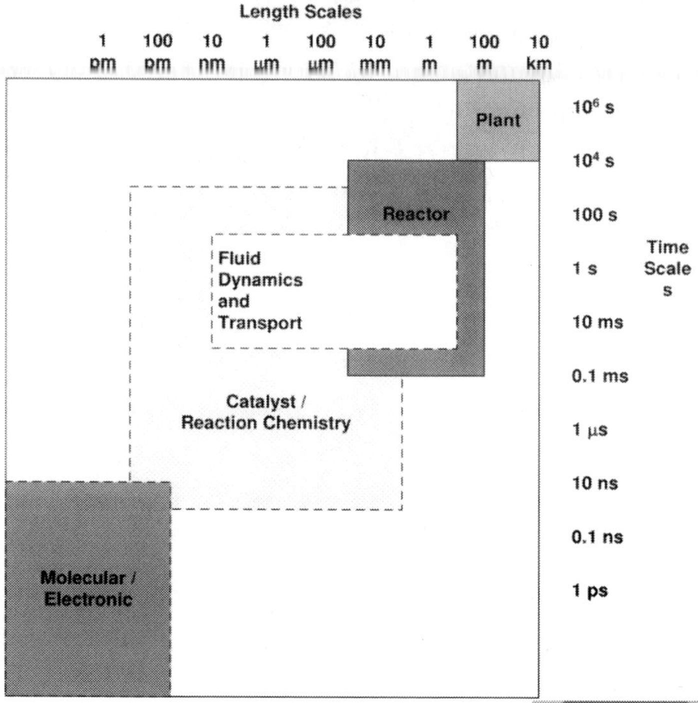

2023 AIChE Annual Meeting – Orlando, FL – Paper 908

51

How did you do on the pre-test?

- In scale-up, what are the three scales of interest?
- Not the lab, pilot plant, and plant.
  - One person's lab is another's production unit.
- The three scales: macroscale, microscale, and mesoscale
  - Macroscale – Scale of the vessel and impeller
  - Microscale – Scale of the molecule, flow is no longer convective
  - Mesoscale – Scale of the fresh feed stream

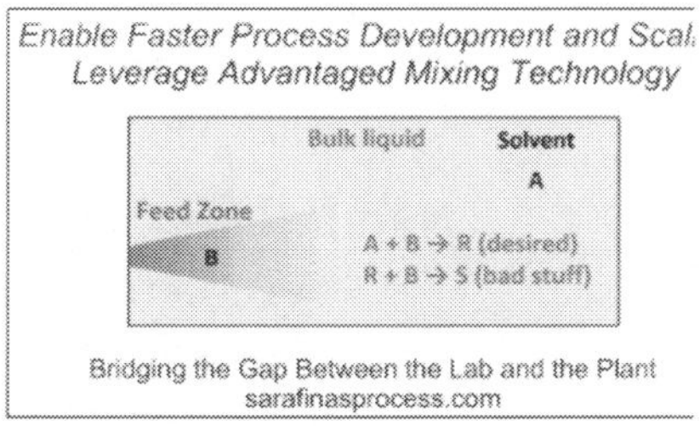

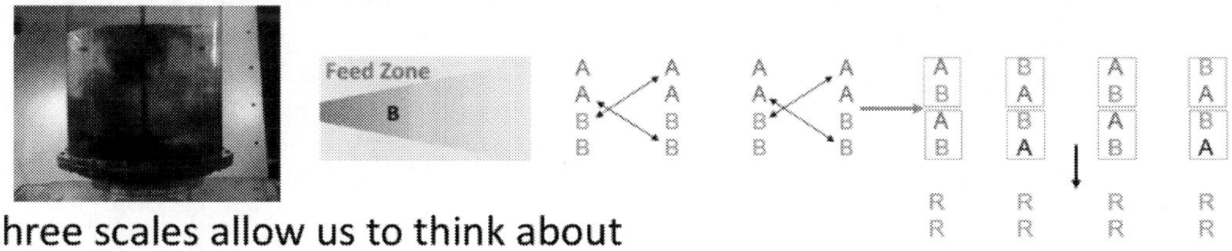

- These three scales allow us to think about mixing rates relative to reaction rates.

This is my favorite quiz that I always give when teaching about competitive rate processes. In chemical reaction scale-up, what are the three scales of interest?
(build) It is not the lab, pilot plant, and plant.
(build) The three scales of interest for scale-up are the macroscale, microscale, and mesoscale.
(build) The macroscale is the size of the vessel or the impeller, and when we talk about macromixing we are referring to our efforts to approach homogeneity in composition or temperature or phase volume fraction across the whole vessel.
(build) The microscale is the scale of the molecule, where flow is no longer convective. Micromixing, in this context, refers to mixing at or near the molecular level.
(build) The mesoscale is the scale of the fresh feed stream, and so mesomixing refers to the gradual disintegration of a feed stream in a turbulent environment down to the length scales where micromixing occurs.
(build) These three scales allow us to consider mixing rates relative to reaction rates.

A look at the Bourne Protocol

- Quick review:
  - Efficient mixing screening through the Bourne Protocol – what it is and why it works
  - Path to optimization by extending into micromixing-mesomixing coordinates
  - Critical characterization of $\varepsilon_{local}/\varepsilon_{mean}$ and u_{local}/V_{tip}
- Needed Enhancements:
  - How best to get and apply local ε and u vs characteristic ε and u
    - How good is "good enough" for screening studies?
    - How do information needs change for troubleshooting and optimization?
    - What is a "surface feed?"
  - More validation cases

2023 AIChE Annual Meeting – Orlando, FL – Paper 90B

53

I'll briefly review the Bourne Protocol as an efficient approach to screening mixing sensitivities in a process. Then I will describe how plotting the results in an operating space that uses micromixing-mesomixing coordinates can be a path to optimization and how it is critical to appropriately characterize the relative energy dissipation and fluid velocity in the feed region.

After that background, I will discuss needed enhancements for screening and optimization studies.

Does Mixing Matter? Prove it doesn't matter…

- From Ed Paul's presentation at the AIChE 2003 Process Development Symposium

Any Rules About Mixing?

- **Always assume there is a mixing problem until proven otherwise**
- **Scale-up will make it worse**
- **Sub-surface addition for reaction and crystallization**
  - **Prove not needed -- not prove that it is**

Reference: E. L. Paul, "Notes on Process Development: Pharmaceutical Intermediates and API's," AIChE Process Development Division Symposium, Pocono Mountain, PA, June 2003.

2023 AIChE Annual Meeting – Orlando, FL – Paper 90B

54

Does mixing matter? Ed Paul said that it always matters, unless you prove otherwise. I will discuss how the Bourne Protocol and the Operating Space are ways to show how mixing matters to your process.

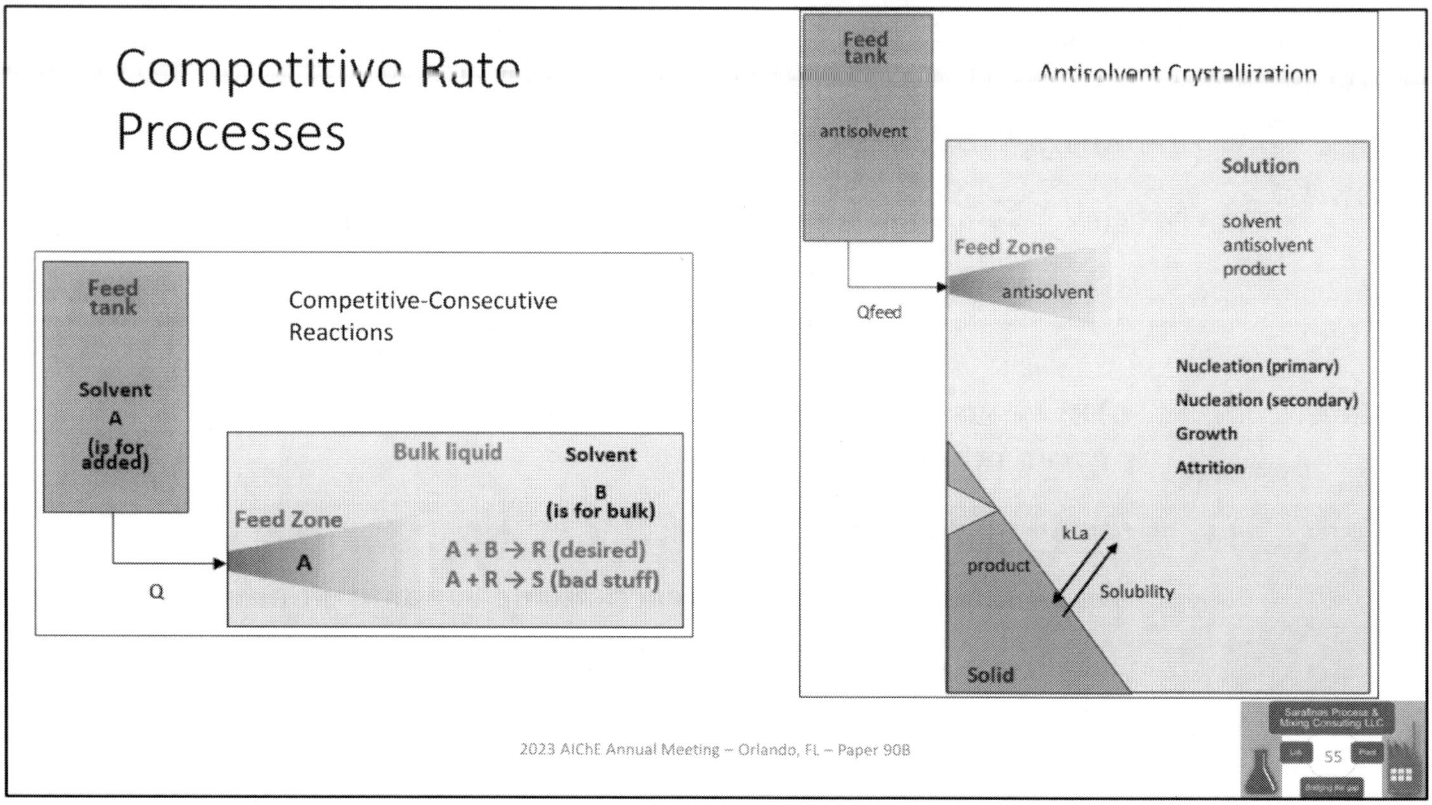

Mixing can matter in competitive rate processes.

I show Process Schemes of two processes with competing rates: a competitive-consecutive reaction on the left and an antisolvent crystallization on the right. These two qualitative chemical and physical models show the phases, components, and rate processes, but the one rate process that I want to emphasize is the disintegration of the feed plume to eliminate the concentration gradients of the added material.

In the Process Scheme on the left, high concentrations of A in the feed plume will produce higher rates of the undesired reaction. In the Process Scheme on the right, high antisolvent concentrations in the feed plume will locally put the system in an undesired part of the phase diagram where the solids may crash out due to high primary nucleation, leading to an unacceptable crystal size distribution or other process problems.

While we may understand reaction kinetics and reaction rate constants, we often do not have information on the rate constants for primary and secondary nucleation, particle growth, and attrition. We need an experimental approach to see how mixing matters in these systems.

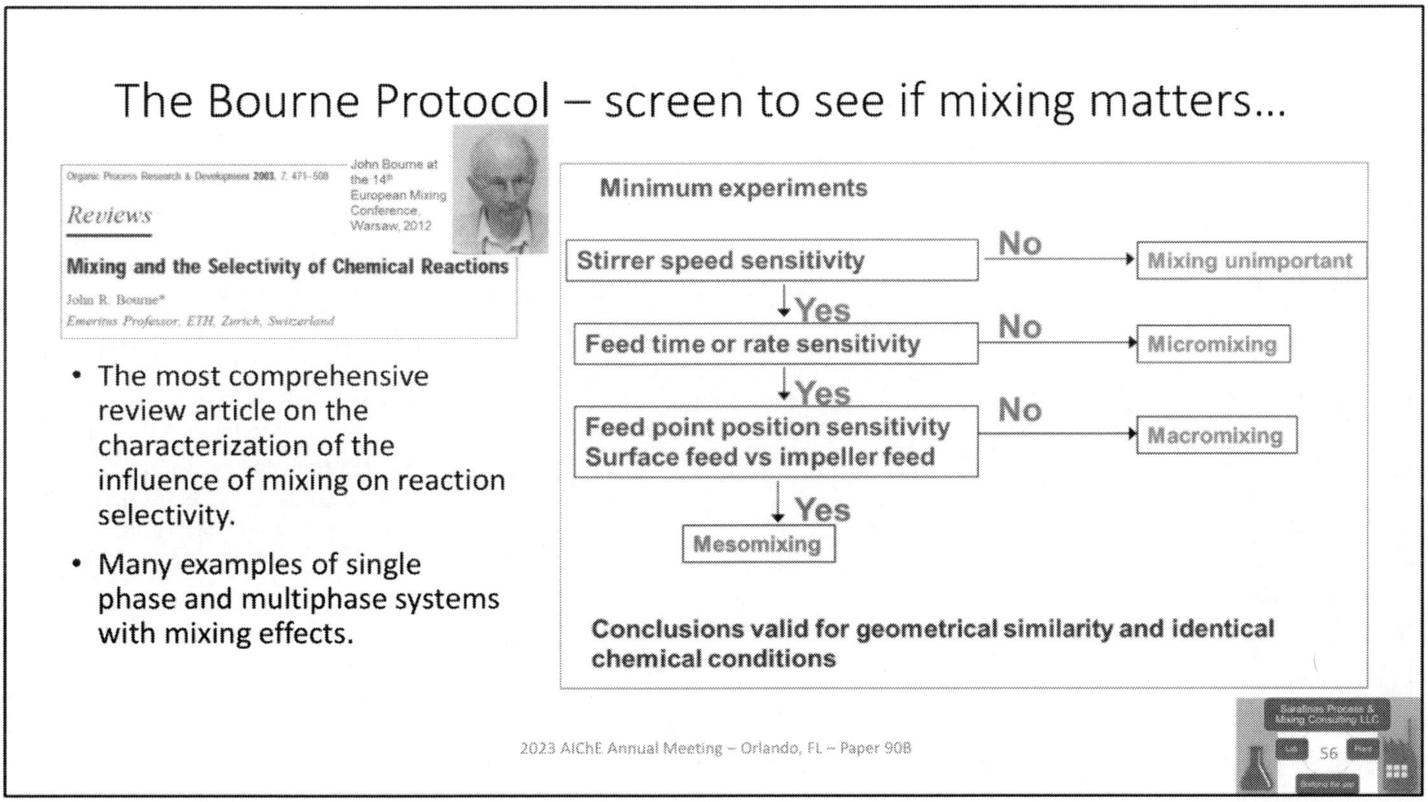

In John Bourne's 2003 review article on Mixing and the Selectivity of Chemical Reactions, he proposes a protocol which we've used successfully to see if mixing matters in semi-batch processes, using the actual process chemistry (not using model reactions). There are three sequential parts to the protocol: stirrer speed sensitivity, feed time or feed rate sensitivity, and feed location sensitivity.

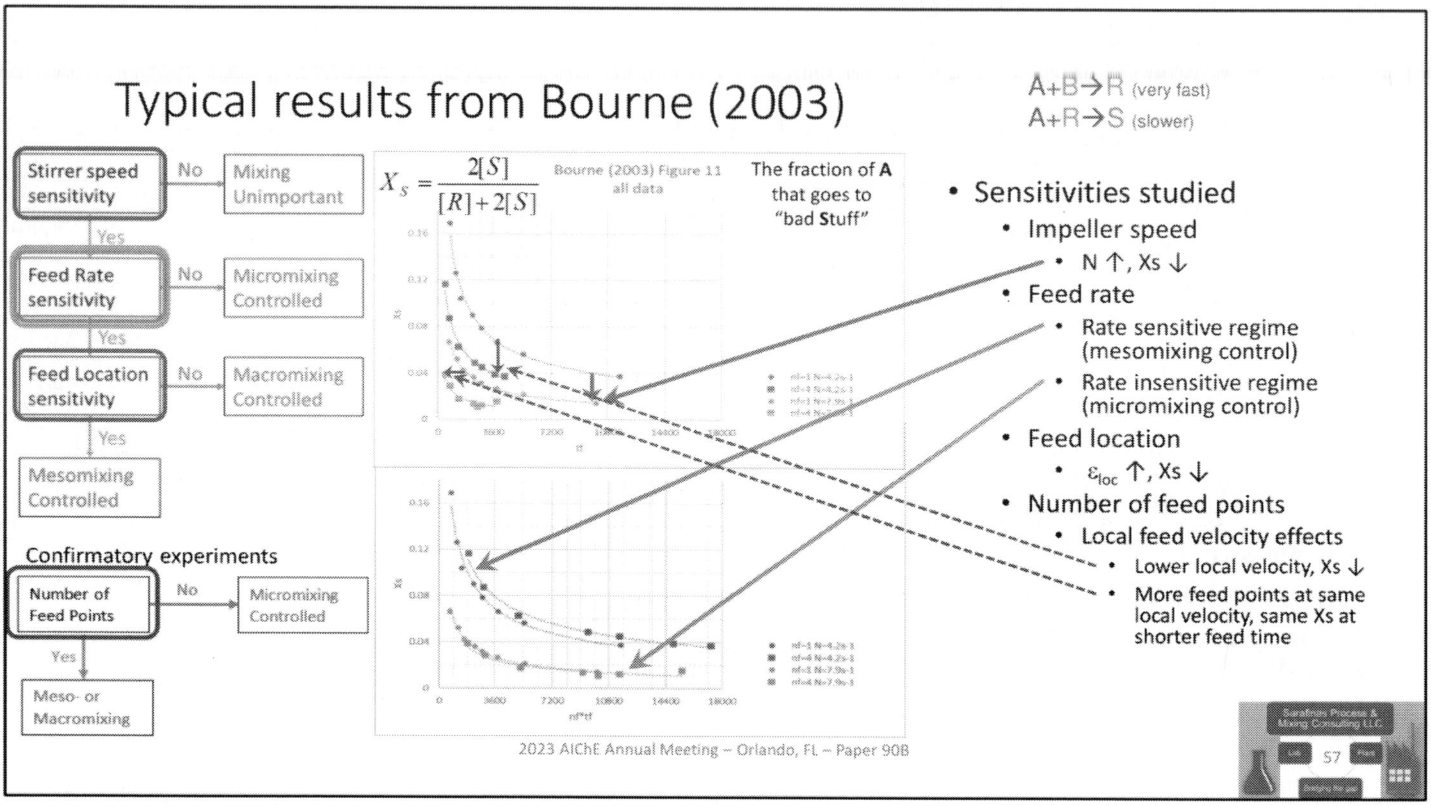

In previous talks I have gone over this slide in much detail, but let me very briefly illustrate how the protocol would be a more efficient means of understanding how mixing matters, using some data from Bourne's 2003 paper. Here is my replot of a couple of figures where Bourne plotted Xs, the fraction of added A that is lost to the undesired S, versus feed time for a large number of experiments at a couple of impeller speeds and a couple of numbers of feed nozzles.

First, the stirrer speed sensitivity. Increasing the impeller speed reduces the fraction of A that goes to S, so we go onto the second step of the protocol...

...where we look at the effect of feed rate or feed time. There is a region where Xs is sensitive to feed rate or feed time, the mesomixing controlled regime, and there is a region where Xs is insensitive to feed rate or feed time, the micromixing controlled regime.

The feed location sensitivity looks at changes in power per unit mass, though the plots on this slide only show higher intensity impeller region feeds. Just like higher impeller speeds give lower Xs, higher local power per unit mass gives lower Xs.

Bourne also described a confirmatory experiment where the number of feed points are varied. In one case, going from 1 to 4 feed points at the same total feed time reduces the local velocity of the feed and reduces Xs in the mesomixing controlled regime. Alternatively, we can go from 1 to 4 feed points with the same local velocity and reduce the feed time by a factor of 4 while maintaining the same yield loss of A to S.

What's going on in this protocol?

- Controlled experiments, balancing time constants
  - Micromixing (engulfment) time

$$t_e = 17\left(\frac{\nu}{\varepsilon_{local}}\right)^{(1/2)}$$

  - Mesomixing time
    - $V_{tip} >> u_{feed}$

$$t_s = 2\left(\frac{d_{feed\_pipe}^{\;2}}{4}\frac{u_{feed}}{u_{local}}\frac{1}{\varepsilon_{local}}\right)^{(1/3)}$$

    - $V_{tip} << u_{feed}$

$$t_s = 2\left(\frac{d_{feed\_pipe}^{\;2}}{\varepsilon_{local}}\right)^{(1/3)}$$

  - Macromixing time
    - Blend time = f(ε_{mean}, fluid properties, geometry)

- Bourne (2003)

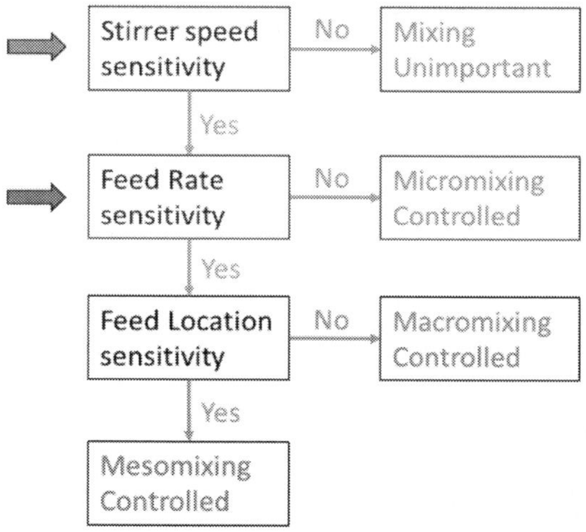

2023 AIChE Annual Meeting – Orlando, FL – Paper 90B

58

This protocol works because it uses controlled experiments to see the effect of different mixing time constants on our process.

When we do the impeller speed sensitivity, all three mixing time constants change. The micromixing (or engulfment) time changes because of the change in the local power per unit mass in the vicinity of the feed. The mesomixing time changes because of the change in local power per unit mass and local velocity in the vicinity of the feed. And the macromixing or blend time changes because the total power per total mass changes.

In the feed rate sensitivity, predominantly the feed velocity drives changes in the mesomixing time constant, though there could be slight changes in some of the local parameters. Bourne said that if there is no feed rate sensitivity, then micromixing controls and to step out of the protocol. I look at it a little differently.

What's going on in this protocol?

- Controlled experiments, balancing time constants
  - Micromixing (engulfment) time

$$t_e = 17\left(\frac{\nu}{\varepsilon_{local}}\right)^{(1/2)}$$

  - Mesomixing time
    - $V_{tip} >> u_{feed}$ $$t_s = 2\left(\frac{d_{feed\_pipe}^{2}}{4}\frac{u_{feed}}{u_{local}}\frac{1}{\varepsilon_{local}}\right)^{(1/3)}$$
    - $V_{tip} << u_{feed}$ $$t_s = 2\left(\frac{d_{feed\_pipe}^{2}}{\varepsilon_{local}}\right)^{(1/3)}$$
  - Macromixing time
    - Blend time = f(ε_{mean}, fluid properties, geometry)

- My revision

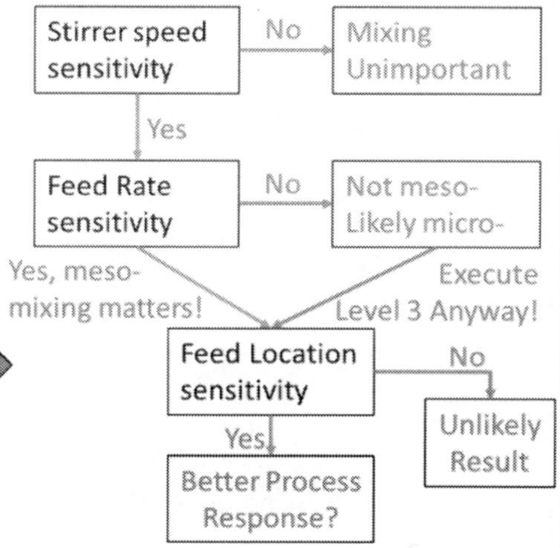

2023 AIChE Annual Meeting – Orlando, FL – Paper 90B

I prefer to run the feed location sensitivity no matter what result I see from the feed rate sensitivity, so I can see if I can get a better process response from a different feed location.

What's going on in this protocol?

- Controlled experiments, balancing time constants
  - Micromixing (engulfment) time

$$t_e = 17\left(\frac{\nu}{\varepsilon_{mean}(\varepsilon_{local}/\varepsilon_{mean})}\right)^{(1/2)}$$

  - Mesomixing time
    - $V_{tip} >> u_{feed}$

$$t_s = 2\left(\frac{d_{feed\_pipe}^{\;2}}{4}\,\frac{u_{feed}/V_{tip}}{u_{local}/V_{tip}}\,\frac{1}{\varepsilon_{mean}(\varepsilon_{local}/\varepsilon_{mean})}\right)^{(1/3)}$$

    - $V_{tip} << u_{feed}$

$$t_s = 2\left(\frac{d_{feed\_pipe}^{\;2}}{\varepsilon_{mean}(\varepsilon_{local}/\varepsilon_{mean})}\right)^{(1/3)}$$

  - Macromixing time
    - Blend time = f(ε_{mean}, fluid properties, geometry)

- My revision

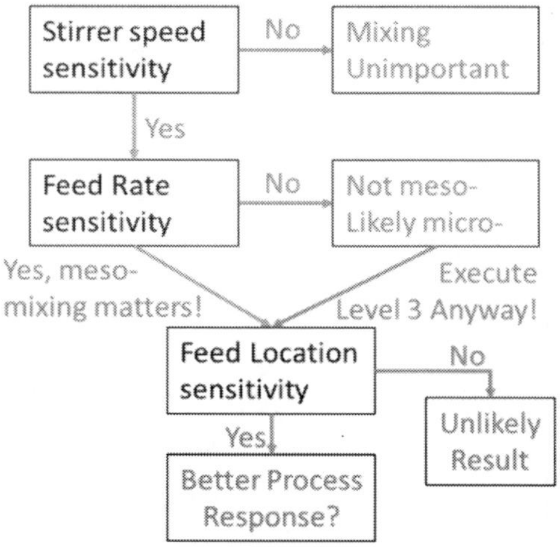

2023 AIChE Annual Meeting – Orlando, FL – Paper 90B

60

The local power per unit mass and local velocity at the feed location can be challenging to characterize when estimating the micromixing and mesomixing times. It is much easier to calculate the tank average power per unit mass and the tip speed. We then need to know the ratio of the local to mean energy dissipation and the ratio of the local fluid velocity to tip speed, which are often characterized from laser-doppler anemometry data or computational fluid dynamics analyses.

Estimates of mixing parameters – a brief word of caution…

- *All* mixing calculations that *any* software provides are *estimates* of mixing parameters.
    - Based on your inputs
        - Geometry
        - Physical properties
    - Based on the validity of the approaches used in the software
        - Engineering correlations based on experimental results
        - Other physical models
    - Use of any software does not allow the user to turn one's brain off
- Use the estimates of mixing parameters for comparative purposes

Now since I specialize in mixing technology, I need to give a word of caution about the estimates of mixing parameters. I don't have time to dive into detail on this, but I want to point out that all mixing calculations that any software provides are estimates of mixing parameters. These estimates are based on your inputs on the vessel and impeller geometry and your knowledge of your phase properties. Consider carefully how your phase properties change during your process and do multiple calculations to understand the impact of those property changes on the estimated mixing parameters. The estimates are also affected by the validity of the approaches used in the software. Some software uses engineering correlations based on experimental results, while other software uses other physical models. Just because "the software said so" is no reason to not think about the reasonableness of the answers and if the models that are being used are appropriate for your particular system.

Today I'll be showing how the mixing parameter estimates should be used for comparative purposes.

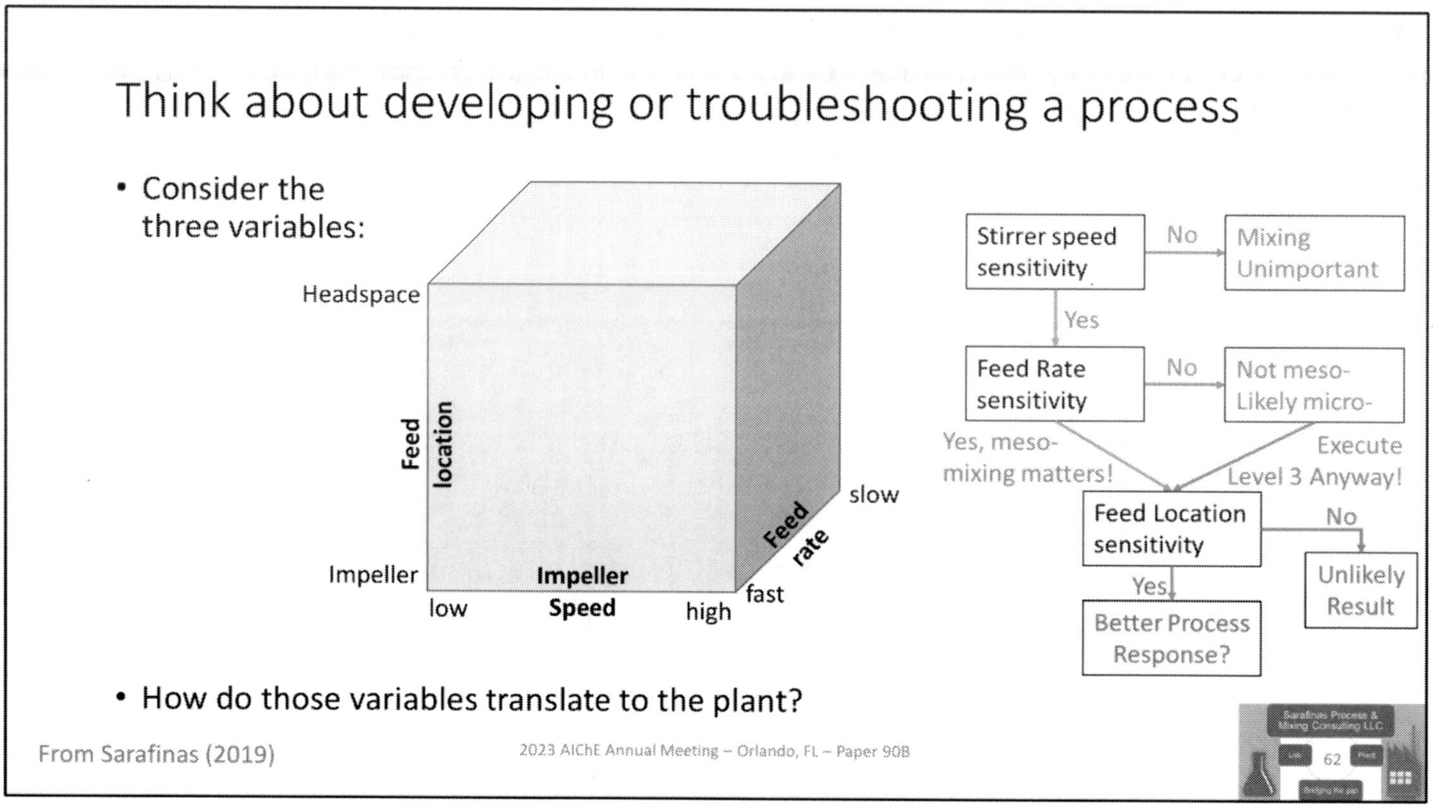

Over the last few years, I have been discussing how we can take the three variables we study in the Bourne Protocol (impeller speed, feed rate, and feed location) and transform those in a way that is translatable to the plant.

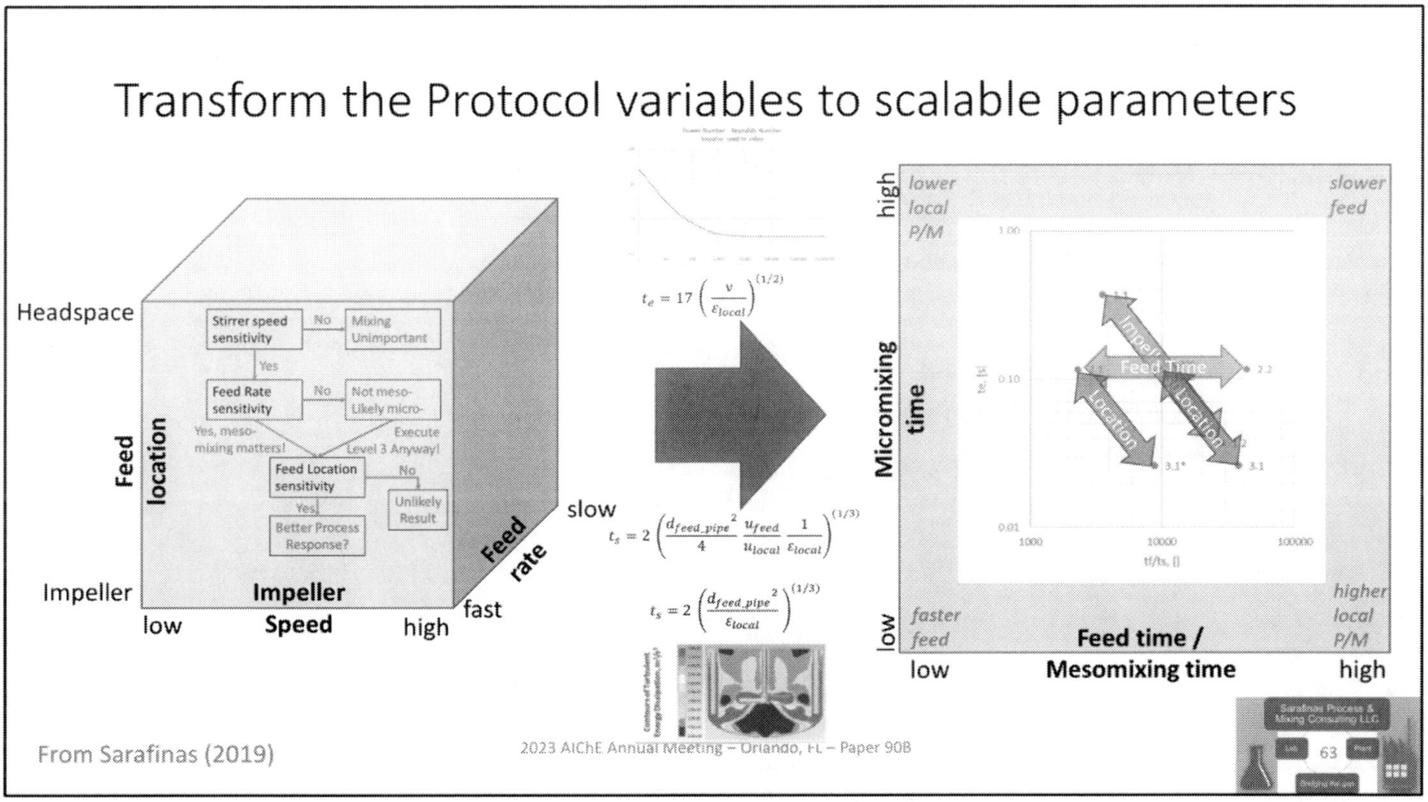

I have shown how we could collapse the three variables of the Bourne Protocol into a scalable operating space looking at micromixing and mesomixing phenomena. The y-axis is micromixing time and the x-axis of feed time/mesomixing time. That was originally based on studies of crystallization phenomena (Torbacke and Rasmusson, 2004) and more recently studies of reaction systems (Brown, 2022) showed the criticality of the feed time / mesomixing time ratio in the scale-up their reactive mixing studies. When we look at the impeller speed and feed location axes in the original Bourne Protocol cube, we're really looking at a range from low power per unit mass (headspace feed with low impeller speeds) to high power per unit mass (impeller region feed with high impeller speeds). We can use the power number – Reynolds number curve and information about local energy dissipation to transform impeller speed and feed location variables through power per unit mass into the operating space using the equations for micromixing and mesomixing times. Feed rate gets transformed using the mesomixing time equations.

Here is the protocol plotted in the micromixing-mesomixing space. The impeller sensitivity covers these points. The feed time sensitivity covers these points. And the feed location sensitivity has a similar effect as impeller speed in these coordinates.

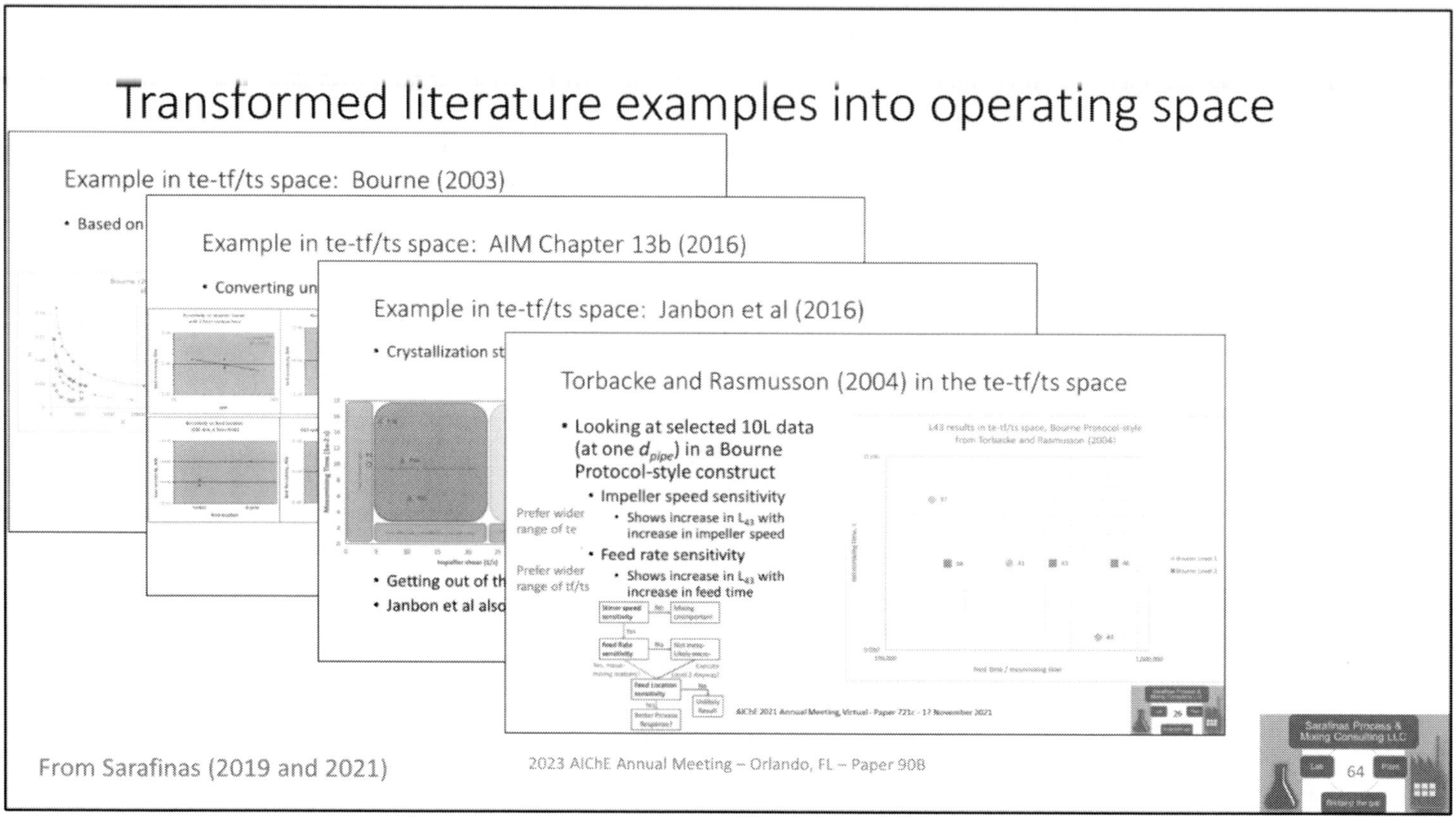

In earlier talks, I showed that a number of cases from the literature were well characterized using this Operating Space approach:

- The data from Bourne's paper that we looked at earlier
- Our chapter from Advances in Industrial Mixing converting an uncontrolled precipitation into a well-behaved reactive crystallization
- A paper from Astra Zeneca on crystallization studies at three production scales, with varying degrees of process problems
- And the Torbacke and Rasmusson paper on reactive crystallization of benzoic acid

Scale from 8000L to 100L – characterize plant vessel

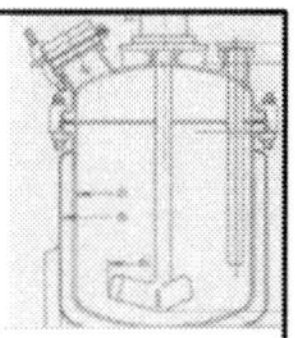

- Scale between two glass-lined vessels:
  - Pfauder AE-6300 (8000L)
  - Pfaudler AE-63 (100L).
- Characterize the plant vessel for operating ranges for headspace feed

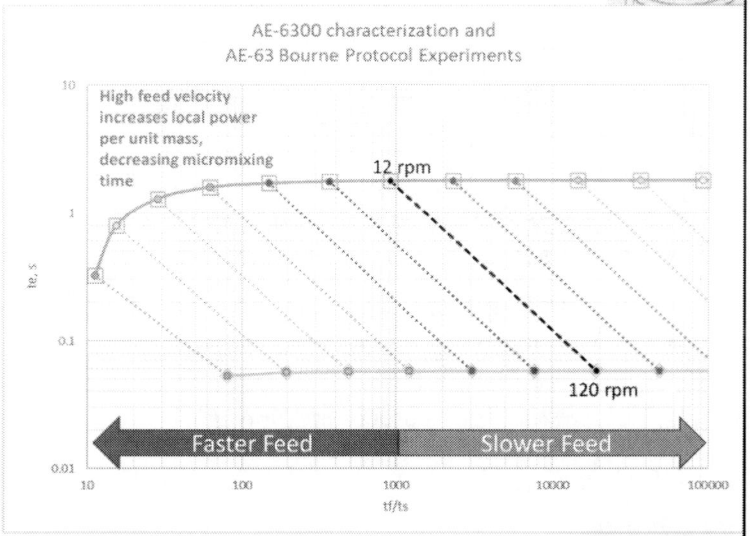

From Sarafinas (2020)

2023 AIChE Annual Meeting – Orlando, FL – Paper 90B

65

We can use the same operating space to characterize a plant scale vessel for possible operating conditions, with tie lines of constant feed time.

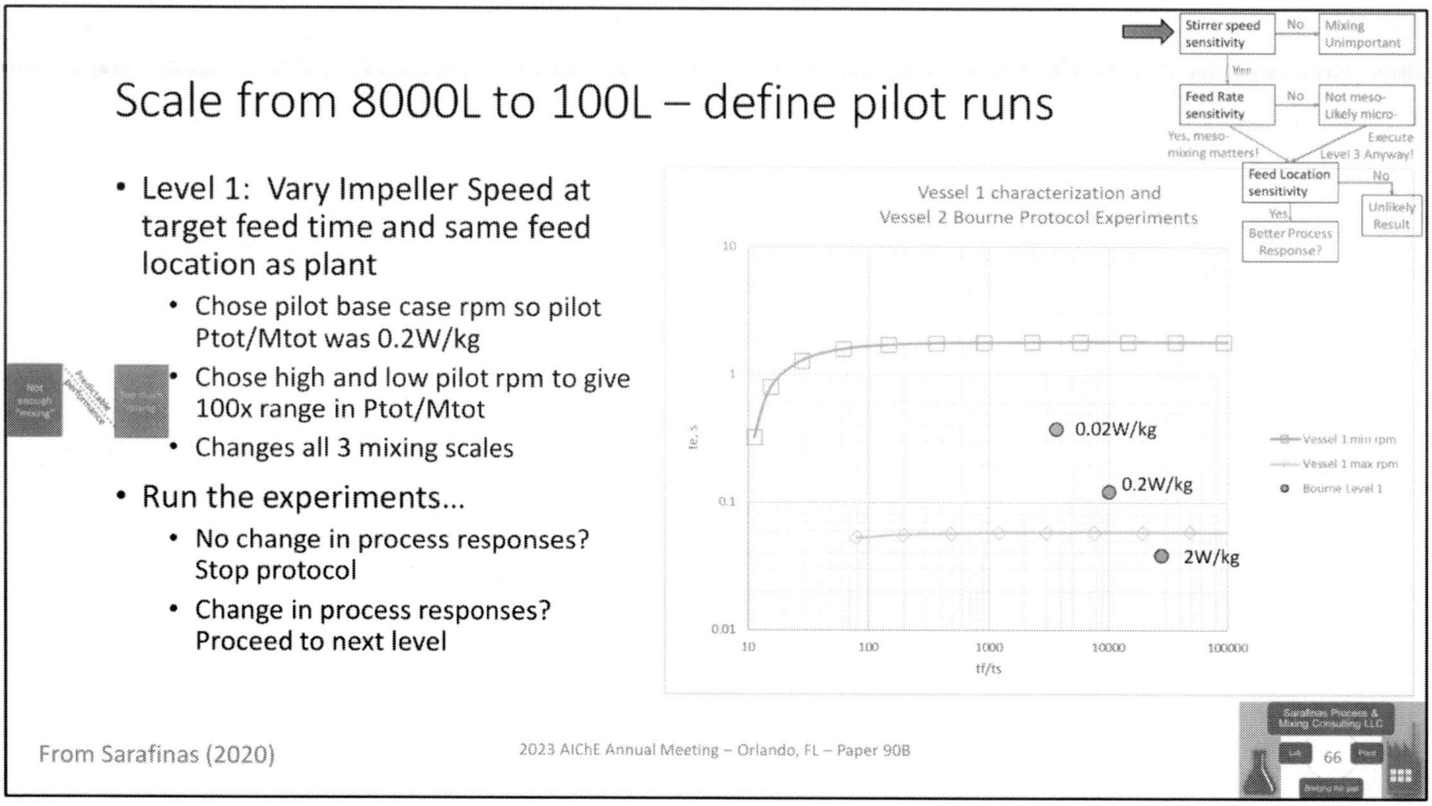

And we can superimpose the Bourne Protocol conditions for a pilot vessel on that operating space, with the impeller speed sensitivity...

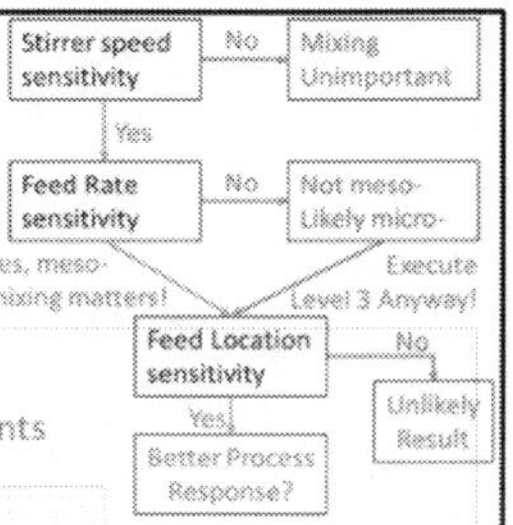

Scale from 8000L to 100L – define pilot runs

- Level 2: Vary Feed Time at base case rpm and feed location
  - Chose pilot base case feed time to match the plant tf/ts, which requires shorter pilot feed time
  - Chose high and low pilot feed time for 9x range
- Run the experiments…
  - No change in process responses? Not mesomixing
  - Change in process responses? Mesomixing matters

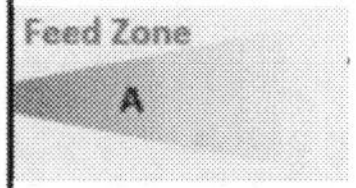

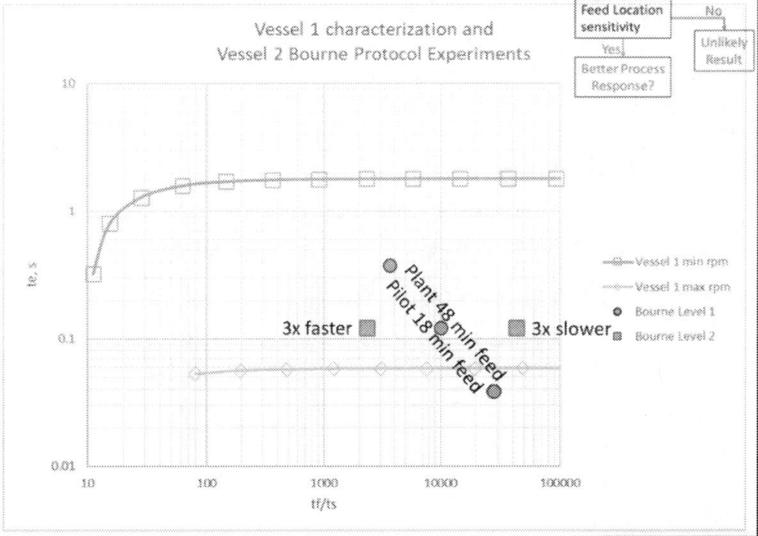

From Sarafinas (2020)

2023 AIChE Annual Meeting – Orlando, FL – Paper 90B

67

…the feed time sensitivity…

Scale from 8000L to 100L – define pilot runs

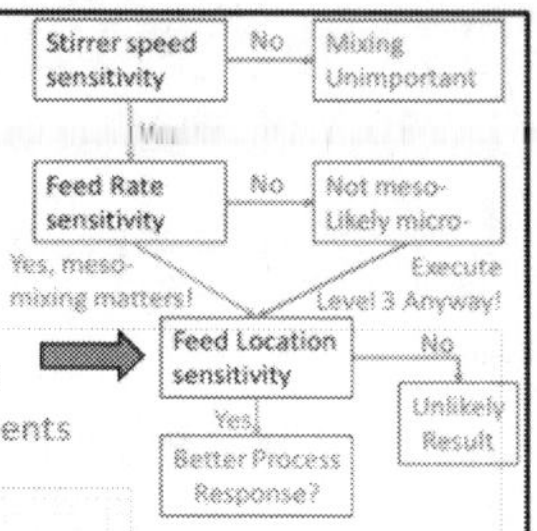

- Level 3: Change the feed location at the base case rpm
  - Moving feed location from headspace to impeller increases P/M which decreases te and increases tf/ts

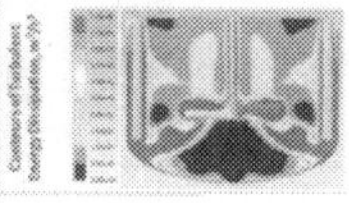

    Can choose feed time for base case or fast feed (lower tf/ts)
  - Watch out for feed pipe backmixing
- Run the experiments…
  - What conditions give the best process response?

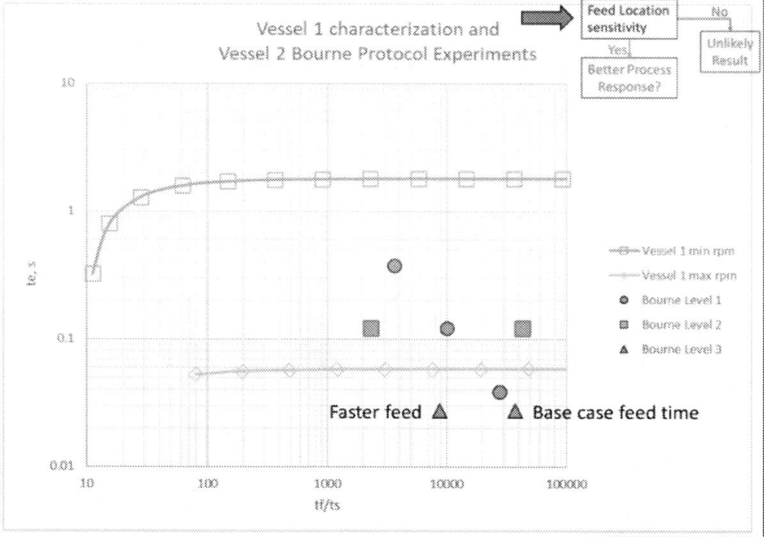

From Sarafinas (2020)

2023 AIChE Annual Meeting – Orlando, FL – Paper 90B

68

…and the feed location sensitivity…all with the intention of finding the regions in the operating space that give us the best process responses which we can scale to the production vessel.

But we need the local parameters…

- Micromixing (engulfment) time

$$t_e = 17\left(\frac{\nu}{\varepsilon_{mean}(\varepsilon_{local}/\varepsilon_{mean})}\right)^{(1/2)}$$

- Mesomixing time
  - $V_{tip} >> u_{feed}$

$$t_s = 2\left(\frac{d_{feed\_pipe}^{\;2}}{4}\,\frac{u_{feed}/V_{tip}}{u_{local}/V_{tip}}\,\frac{1}{\varepsilon_{mean}(\varepsilon_{local}/\varepsilon_{mean})}\right)^{(1/3)}$$

From Sarafinas (2021)

2023 AIChE Annual Meeting – Orlando, FL – Paper 90B

69

But remember, to make that transformation we need good understanding of the local parameters (energy dissipation and velocity in the bulk fluid in the vicinity of the feed) relative to the easily calculable vessel parameters (mean energy dissipation and tip speed).

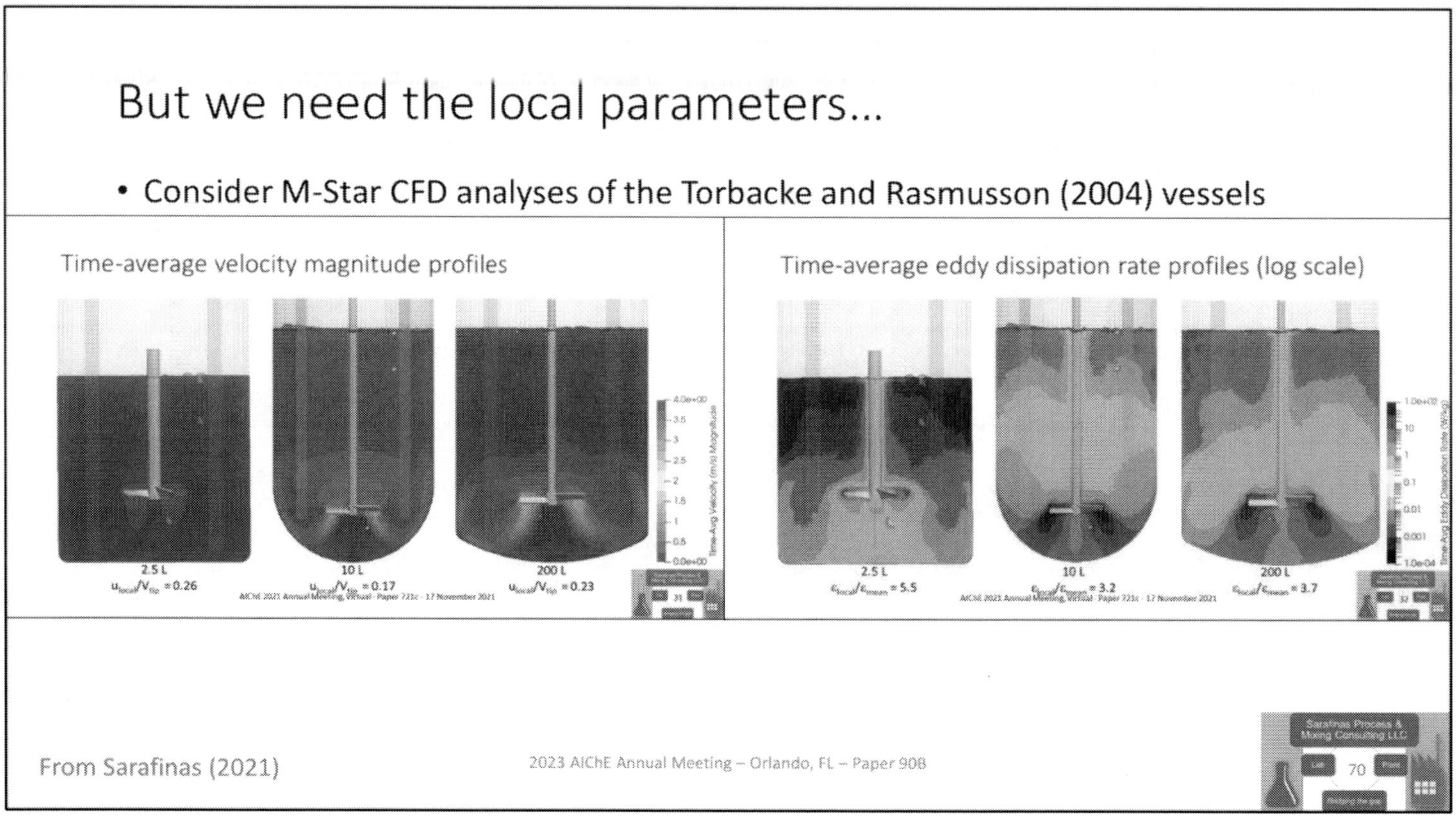

At last fall's AIChE Annual Meeting, I discussed some CFD analyses that I made using M-Star on the Torbacke and Rasmusson experimental equipment and conditions and compared the CFD result to three other approaches to obtain the local parameters. The M-Star CFD results for their specific equipment and feed locations were quite close to the heuristics used in the DynoChem Mixing and Heat Transfer Toolbox,

But we need the local parameters…

- Consider M-Star CFD analyses of the Torbacke and Rasmusson (2004) vessels

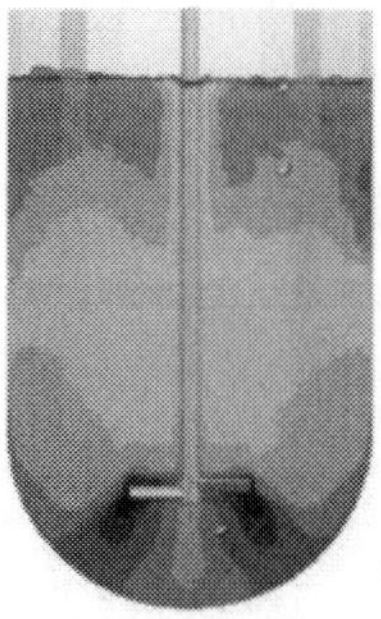

Comparison of $\varepsilon_{local}/\varepsilon_{mean}$ and u_{local}/V_{tip} methods

- Consider 10L at end-of-feed and 900rpm in the impeller region

	$\varepsilon_{local}/\varepsilon_{mean}$	u_{local}/V_{tip}
Default (DynoChem MHTT)	2	0.2
$\varepsilon_{max}/\varepsilon_{mean}$	262	
"30% of power dissipated … impeller swept volume"	71	
M-Star simulations	3.2	0.17

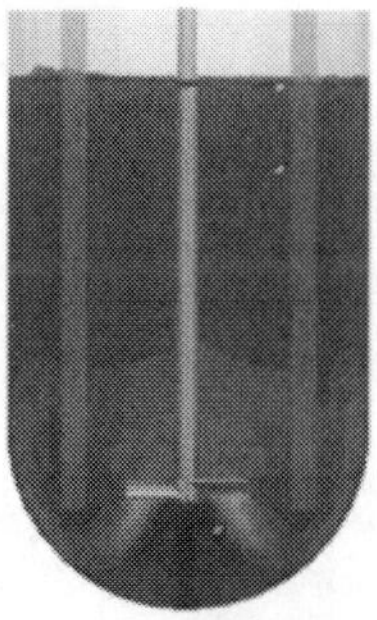

- Default is not too far from simulation, but slight change in feed location could increase factor greatly
- The feed may not see ε_{max} or the "30%" region
- Very slight impact on operating space plots for Torbacke and Rasmusson study AIChE 2021 Annual Meeting, Virtual - Paper 721c - 17 November 2021

From Sarafinas (2021)

2023 AIChE Annual Meeting – Orlando, FL – Paper 90B

The M-Star CFD results for their specific equipment and feed locations were quite close to the heuristics used in the DynoChem Mixing and Heat Transfer Toolbox, but some slight deviations in feed locations could have made large changes in the local power per unit mass and the local velocity. That change would mean that the default values would have given inappropriate micromixing and mesomixing times, leading to incorrect locations in the Operating Space.

But we need the local parameters…

- Consider M-Star CFD analyses of the Torbacke and Rasmusson (2004) vessels

Need for better $\varepsilon_{local}/\varepsilon_{mean}$ and u_{local}/V_{tip}

- Jordan et al (2020) – Bristol Myers Squibb
  - Reverse addition reactive crystallization
  - Used Bourne Protocol and transformed into operating space
  - Observed improved interpretation of results in operating space using CFD to characterize $\varepsilon_{local}/\varepsilon_{mean}$ and u_{local}/V_{tip}
- Further reinforces the need for improved estimates of $\varepsilon_{local}/\varepsilon_{mean}$ and u_{local}/V_{tip} for best application of the Bourne Protocol results for optimization in the operating space.

AIChE 2021 Annual Meeting, Virtual - Paper 721c - 17 November 2021

From Sarafinas (2021)

2023 AIChE Annual Meeting – Orlando, FL – Paper 90B

This same effect was described in at the 2020 Annual Meeting in the paper by Bristol-Myers Squibb (Jordan et al 2020). Additionally, two approaches to the maximum and "impeller region" energy dissipation gave much higher results than the process fluid would have seen in the vicinity of the feed.

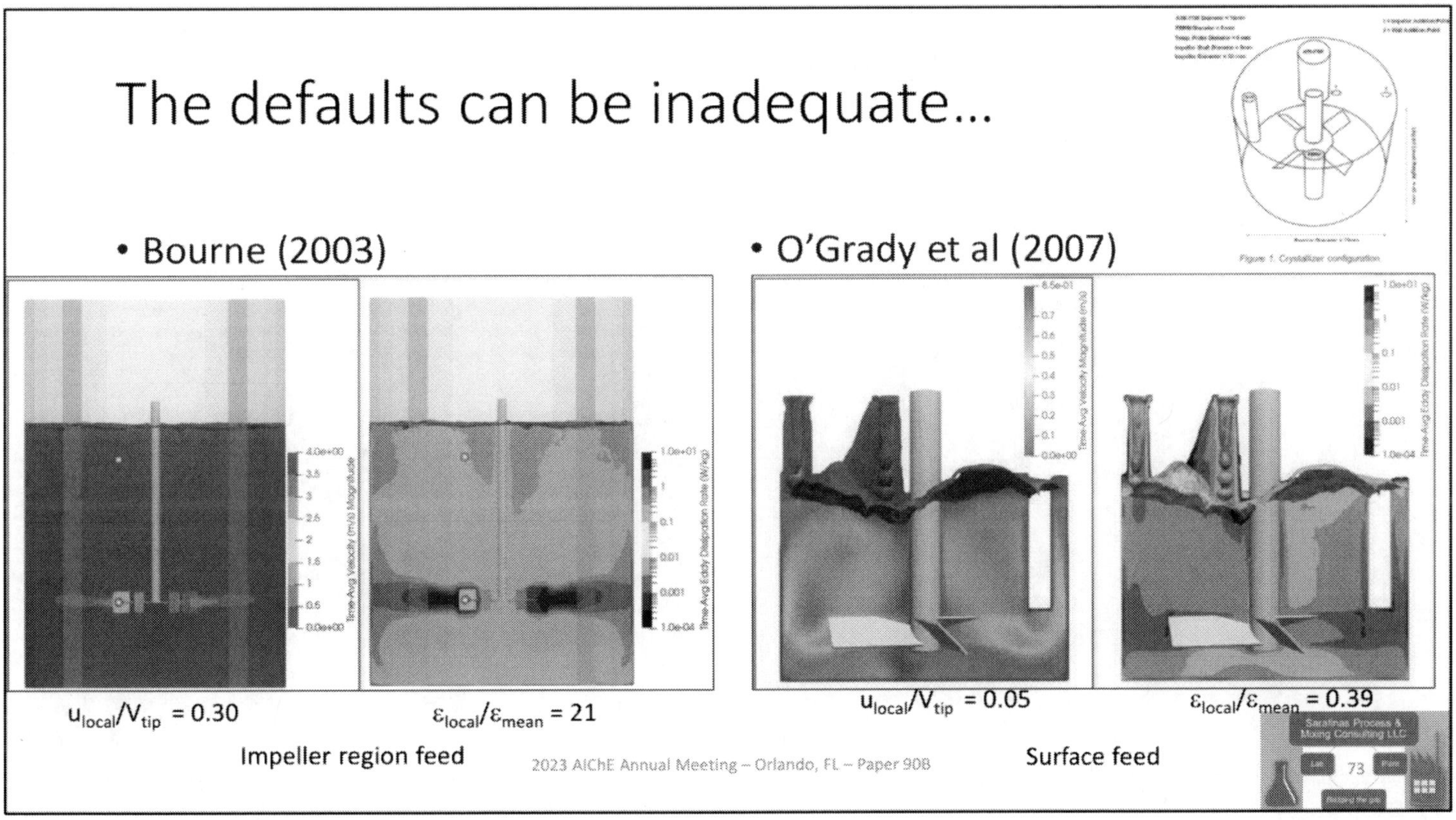

Here are my CFD analyses of equipment and operating conditions from two papers, the Bourne and Hilber configuration shown in Bourne (2003) and the antisolvent crystallization study by O'Grady (2007). The default parameters for the impeller region and the surface feed do not apply well to these systems. Bourne used a feed position directly in the discharge of the radial disk turbine, while O'Grady had a surface feed in a vessel that was partially baffled by cylindrical probes. Understanding the local energy dissipation and the local velocity is critical for appropriately analyzing these systems.

Wait…what do you mean by "surface feed?"

- Bourne and Hilber (1990)

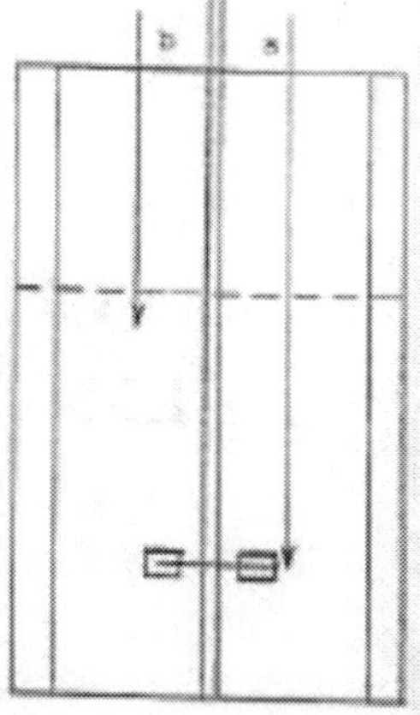

Figure 2. Semi-batch stirred tank reactor with Rushton turbine and (a) feed position turbine, (b) feed position upper.

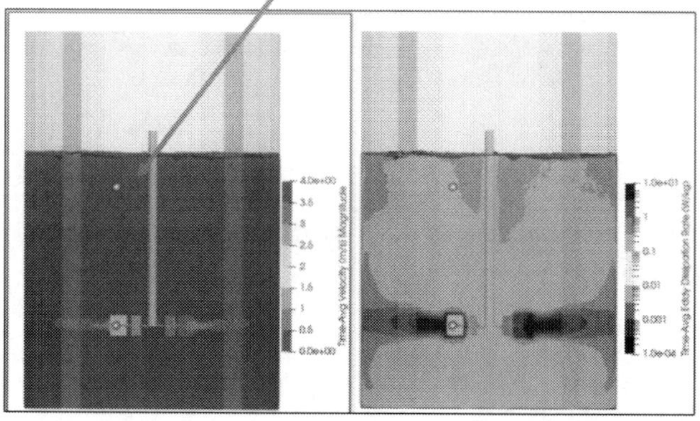

2023 AIChE Annual Meeting – Orlando, FL – Paper 908

"Upper" feed position in Bourne's work is not a true feed to the surface, but slightly subsurface.

I don't know if this is real or not…

…but it sure looks… interesting?

This is a quick screening model using M-Star of a "slow" (dripping) feed to a partially baffled cylindrical lab flask agitated with a 45PBT4 to see what I see, not a careful analysis of the increase in ε_{local} when a drop hits the surface.

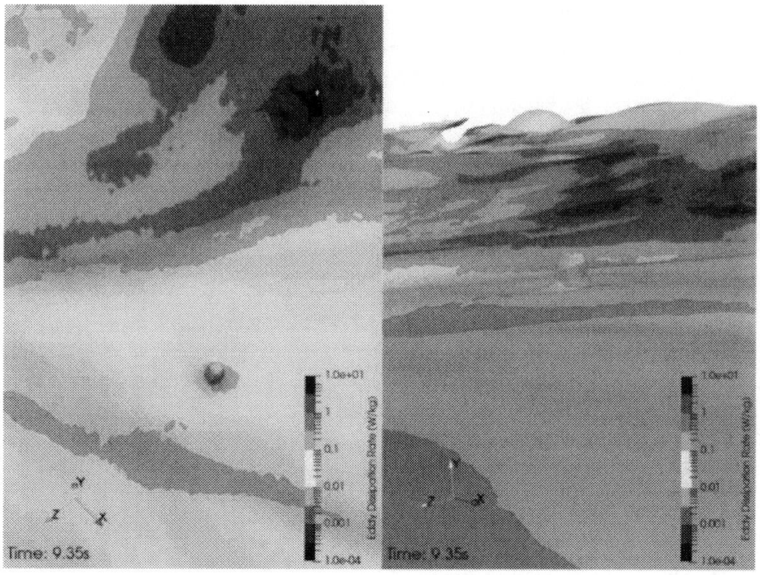

Has anyone seen any validation of energy dissipation in the vicinity of a plunging jet or dripping feed?

Needed Enhancements

- Knowledge base for $\varepsilon_{local}/\varepsilon_{mean}$ and u_{local}/V_{tip}
    - For now, I'm using M-Star CFD to characterize all local (feed region) parameters for transformation to the micromixing-mesomixing operating space
    - How good is "good enough" for a screening study? For optimization?
    - Plan to continue simulations and consider 2nd or 3rd or 4th Bourne reaction
- Further validation of Bourne Protocol and Operating Space approach
    - Scalability:
        - More results at various scales
        - "Good" and "bad neighborhoods" for processes
    - Different competitive rate systems
        - Reactions
        - Multiphase systems
        - Phase-change systems
        - Concentration sensitivity

2023 AIChE Annual Meeting – Orlando, FL – Paper 90B

76

The key element of driving optimization out of the Bourne Protocol is knowledge of the ratios e_local/e_mean and u_local/Vtip. The factorized approach for the ratio of local to average power per unit mass and ratio of local velocity to tip speed needs more detailed information to be effective. Currently, I am using M-Star CFD to characterize all local parameters prior to transforming to the micromixing-mesomixing Operating Space. I plan to continue to investigate how good our estimates need to be for the screening study and for optimization. I'll continue simulations and consider some with the 2nd or 3rd or 4th Bourne reactions, perhaps looking at the time dependent behavior in the feed plume. I may also investigate how well the CFD agrees with Schoenmaker's method for updating the effect of feed rate on local energy dissipation.

We need to look at more validation cases for both the Bourne Protocol and for the Operating Space approach. I'd like to hear more results at various scales, looking at how well the method predicts where your process does well and perhaps more importantly, where it fails.

Some different competitive rate processes for us to consider are competitive-consecutive and competitive reaction systems, multiphase systems, and phase change systems, as well as systems where different reagent concentrations are used.

Summary

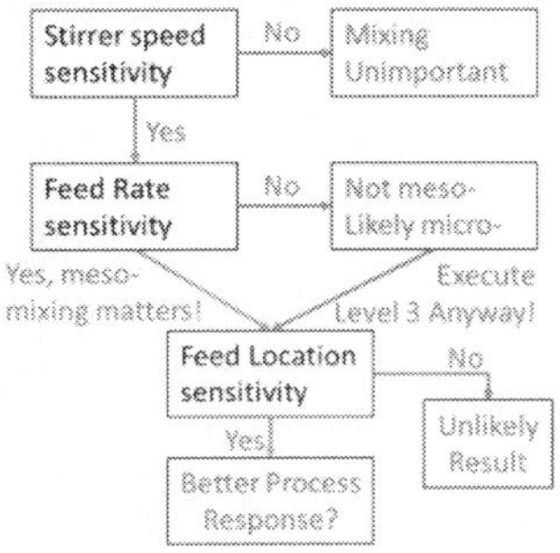

- See if mixing matters:
  Use the Bourne Protocol to screen mixing effects in systems with potential competitive rate processes.
- Consider this scalability space:
  Transform the Bourne Protocol factor settings into the scaled te-tf/ts space to assess acceptable and unacceptable operating regions
- Use reliable $\varepsilon_{local}/\varepsilon_{mean}$ and u_{local}/V_{tip}
- Screening study first, then focus
  - Does mixing matter?
  - Where are my best process responses?
  - Refine my operating conditions
  - Validate on scale-up

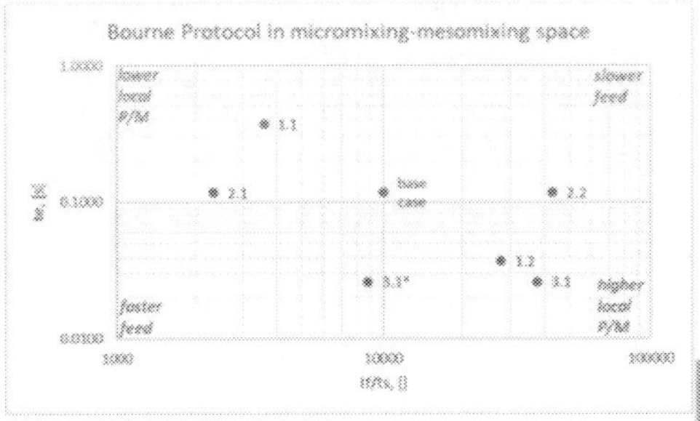

2023 AIChE Annual Meeting – Orlando, FL – Paper 90B

77

To summarize, I recommend that we

- Use the Bourne Protocol to screen mixing effects to see if mixing matters in systems with potential competitive rate processes.
- Consider this scalability space, by transforming the Bourne Protocol factor settings into the te-tf/ts space to assess acceptable and unacceptable operating regions.
- Use reliable estimates of local energy dissipation and local velocity when transforming to the Operating Space .
- Screen first, then focus. Use the Bourne Protocol as a screening study to see if mixing matters, then use the te-tf/ts space to see where our best process responses exist. Then we can refine our operating conditions and validate optimal performance on scale-up.

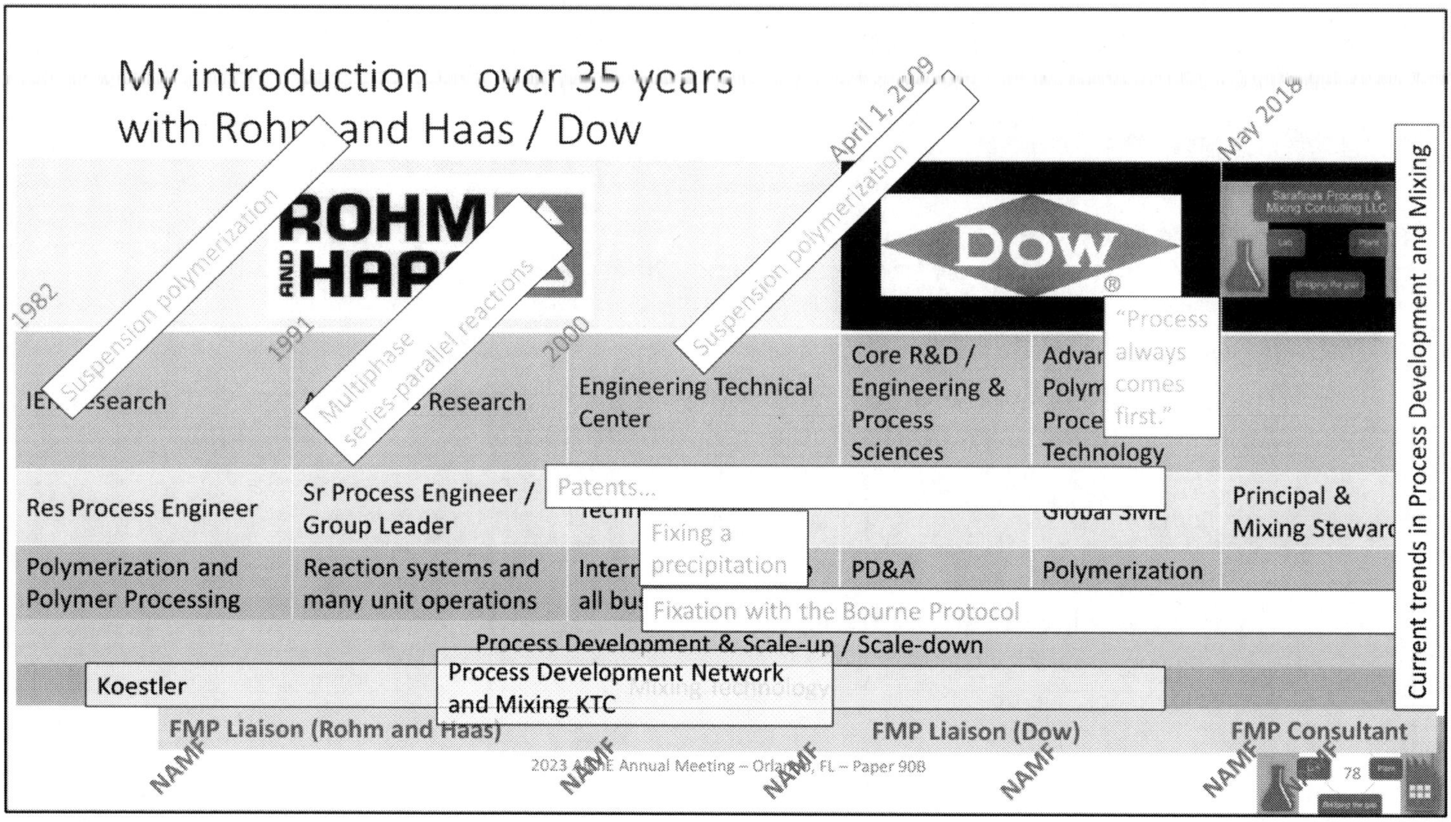
My introduction over 35 years with Rohm and Haas / Dow
ROHM AND HAAS
DOW
1982
1991
2000
April 1, 2009
May 2018
Suspension polymerization
Multiphase series-parallel reactions
Suspension polymerization
"Process always comes first."
Sarafinas Process & Mixing Consulting LLC
Current trends in Process Development and Mixing
IE Research
Research
Engineering Technical Center
Core R&D / Engineering & Process Sciences
Technology
Res Process Engineer
Sr Process Engineer / Group Leader
Patents...
Fixing a precipitation
Principal & Mixing Steward
Polymerization and Polymer Processing
Reaction systems and many unit operations
PD&A
Polymerization
Fixation with the Bourne Protocol
Process Development & Scale-up / Scale-down
Koestler
Process Development Network and Mixing KTC
FMP Liaison (Rohm and Haas)
FMP Liaison (Dow)
FMP Consultant
2023 AIChE Annual Meeting – Orlando, FL – Paper 908
NAMF
78

Where are the process industries going now?

CHEMICAL REVIEWS

The Hitchhiker's Guide to Flow Chemistry

Matthew B. Plutschack, Bartholomäus Pieber, Kerry Gilmore, and Peter H. Seeberger

- What's "trending" in the process industries
  - "Flow Chemistry"
    - Plutschacket al (2017), a paper cited by over 1200 other papers

Flow chemistry involves the use of channels or tubing to conduct a reaction in a continuous stream rather than in a flask. Flow equipment provides chemists with unique control over reaction parameters enhancing reactivity or in some cases enabling new reactions. **This relatively young technology** has received a remarkable amount of attention in the past decade with many reports on what can be done in flow. Until recently, however, the question, "Should we do this in flow?" has merely been an afterthought. This review introduces readers to the basic principles and fundamentals of flow chemistry and critically discusses recent flow chemistry accounts.

Saraiinas Process & Mixing Consulting LLC 79

Where are the process industries going now?

- What's "trending" in the process industries
  - "Flow Chemistry"
    - Plutschacket al (2017), a paper cited by over 1200 other papers
    - Heise et al (2023) on continuous bioprocessing

BIOPROCESSING

CONTINUOUS BIOPROCESSING - THE JOURNEY HAS STARTED AND THE DESTINATION IS CLEARER THAN EVER

ASHLEY REEDER[1], BOB LENICH[2], BRUCE KANE[3], CHARLES HEISE[4*], DONALD GIBBS[5], GRAEME MOODY[6], KRISTINA PLEITT[7], LARA FERNANDEZ-CEREZO[8], MARK SCHOFIELD[9], ANDREA JOHNSTON[6]

*Corresponding author

1. GSK, King of Prussia, USA
2. Emerson, Austin, USA
3. Rockwell Automation, Wisconsin, USA
4. FUJIFILM Diosynth Biotechnologies, Cleveland, United Kingdom
5. Asahi Kasai Bioprocess, Chelmsford, USA
6. BioPhorum Ltd, The Gridiron Building, One Pancras Square, London, United Kingdom
7. Thermo Fisher Scientific, St Louis, USA
8. Merck & Co, Rahway, USA
9. Cytiva, Westborough , USA

Where are the process industries going now?

- What's "trending" in the process industries
  - "Flow Chemistry"
    - Plutschacket al (2017), a paper cited by over 1200 other p
    - Heise et al (2023) on continuous bioprocessing
  - Process Intensification
    - Cheryl Teich CEP article (September 2023) on AIChE RAPID RACER AMPD

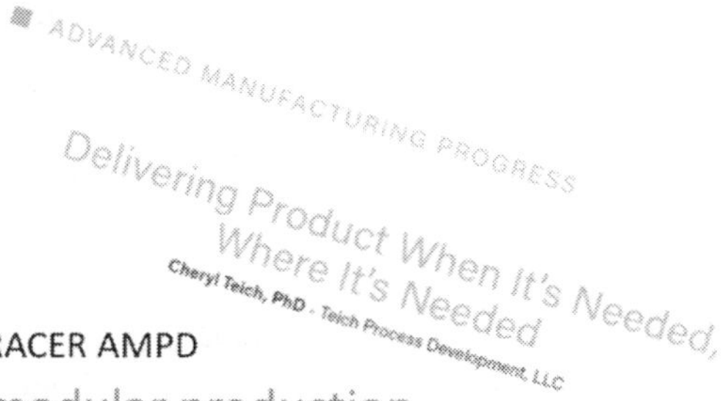

RAPID RACER AMPD has three goals to help localized modular production close this gap (extended shipping times and reliance on remote suppliers):

1. Develop the **accelerated** process development framework
2. Design and operate the **modular process development** CoE at AVN Corp., which will serve as a national asset test-bed for future coronavirus response and a job creation hub in the underserved Appalachian region.
3. Demonstrate the accelerated process development framework for an API precursor specialty chemical that is currently produced offshore

2023 AIChE Annual Meeting – Orlando, FL – Paper 90B

81

RAPID = Rapid Advancement in Process Intensification Deployment manufacturing institute
RACER = Rapid Assistance for Coronavirus Economic Response
AMPD = Accelerated Modular Process Development

Link: https://www.aiche.org/resources/publications/cep/2023/september/advanced-manufacturing-progress-delivering-product-when-its-needed-where-its-needed

Where are the process industries going now?

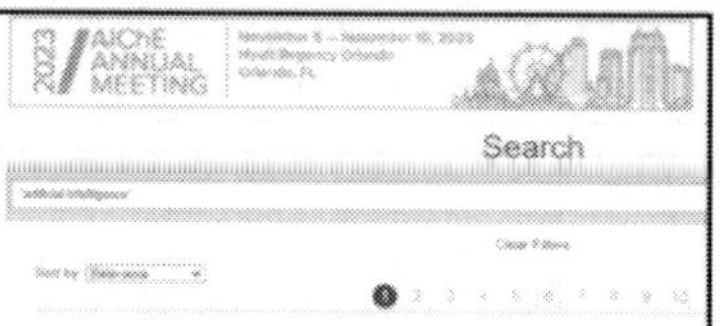

Search on AI gets 100 hits at AIChE 2023 Annual Meeting

- What's "trending" in the process industries
  - "Flow Chemistry"
    - Plutschacket al (2017), a paper cited by over 1200
    - Heise et al (2023) on continuous bioprocessing
  - Process Intensification
    - Cheryl Teich CEP article (September 2023) on AIChE RAPID RACER
  - Artificial Intelligence
    - Bortz et al (2023) on EU KEEN project
    - Chiang et al (2022) on Industry 4.0

Chemie Ingenieur Technik

Review Article

AI in Process Industries – Current Status and Future Prospects

Michael Bortz, Kai Dadhe, Sebastian Engell, Vanessa Gepert, Norbert Kockmann*, Ralph Müller-Pfefferkorn, Thorsten Schindler, and Leon Urbas

DOI: 10.1002/cite.202200247

KEEN has focused on three application areas of AI methods in the process industries: the support of the development of processes, the engineering phase of plant design and development, and the control, monitoring, and optimization of running processes (real-time application).

Where are the process industries going now?

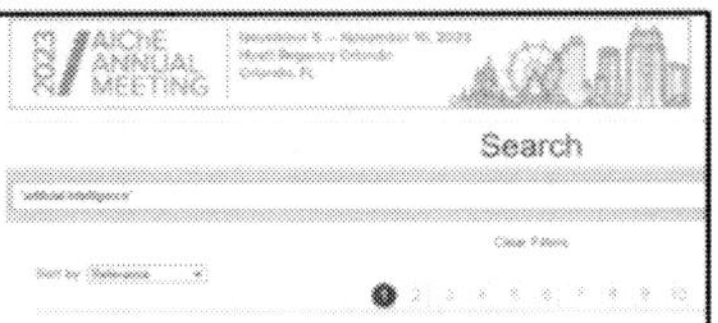

Search on AI gets 100 hits at AIChE 2023 Annual Meeting

- What's "trending" in the process industries
  - "Flow Chemistry"
    - Plutschacket al (2017), a paper cited by over 12(
    - Heise et al (2023) on continuous bioprocessing
  - Process Intensification
    - Cheryl Teich CEP article (September 2023) on AIChE RAPID I
  - Artificial Intelligence
    - Bortz et al (2023) on EU KEEN project
    - Chiang et al (2022) on Industry 4.0

AIChE JOURNAL

RESEARCH ARTICLE

AIChE

Towards artificial intelligence at scale in the chemical industry

Leo H. Chiang, Birgit Braun, Zhenyu Wang, Ivan Castillo

First published: 13 February 2022 | https://doi.org/10.1002/aic.17644 | Citations: 4

In the Industry 4.0 era, the chemical industry is embracing broad adoption of artificial intelligence (AI) and machine learning (ML) methods. This article provides a holistic view of how the industry is transforming digitally towards AI at scale. First, a historical perspective on how the industry used AI to aid humans in better decision-making is shown. Then state-of-the-art AI research addressing industrial needs on reliability and safety, process optimization, supply chain, material discovery, and reaction engineering is highlighted. Finally, a vision of the plant of the future is illustrated with critical components of AI-ready culture, model life cycle management, and renewed role of humans in chemical manufacturing.

83

Where are the process industries going now?

- What's "trending" in the process industries
  - "Flow Chemistry"
    - Plutschacket al (2017), a paper cited by over 1200 other papers
    - Heise et al (2023) on continuous bioprocessing
  - Process Intensification
    - Cheryl Teich CEP article (September 2023) on AIChE RAPID RACER AMPD
  - Artificial Intelligence
    - Bortz et al (2023) on EU KEEN project
    - Chiang et al (2022) on Industry 4.0
- What are the opportunities in these areas for process development and mixing?
  - Improved time to market? Less energy intensity? Safer? Impact on engineers and scientists?
  - PI is an area within AIChE PDD – is greater collaboration needed with NAMF?
  - How are NAMF and PDD responding to the "relatively young technology" of "flow chemistry?"
  - How far are we from a "Process & Mixing AI"? And how do we fill "wisdom" gaps?

Wisdom
Understanding
Knowledge
Awareness
"not a clue"

2023 AIChE Annual Meeting – Orlando, FL – Paper 908

84

Over the last hour…

- Introduction
- Suspension polymerization got me addicted to mixing
- Networking as an effective way of gathering, managing, and disseminating technology within a company

You must network to succeed

- 3 quick examples of how process always comes first
- Ways to efficiently characterize how mixing affects competitive chemical and physical processes: the Bourne Protocol and more
  - Fixing a precipitation leads to a fixation with the Bourne Protocol

You don't need a scale model, you need a model that scales

- Current trends in process development and mixing
- Final thoughts

Success = Process * Skills * Attitude

Understand your process, understand your process, understand your process

Thank you for this honor

2022 – Eric Janz, University of Dayton and Engineered Analysis and Results, LLC
2021 – Robert M. Schisla, Jr., Eastman Chemical Company
2020 – Dr. Richard Cope, Lightning Mixers
2019 – Dr. François Bertrand, Polytechnique Montréal
2018 – Dr. Enrique Gallindo, National Autonomous University in Mexico
2017 – Dr. Richard Grenville, Philadelphia Mixing Solutions
2016 – Prof. Rodney O Fox, Chemical and Biological Engineering, Iowa State University
2015 – null/skip year
2014 – Minye Liu, DuPont Engineering Technology
2013 – Dr. Gary K. Patterson, Missouri University of Science and Technology
2012 – Dr. Julian Fasano, Chemineer (ret.)
2011 – Dr. Roy Penney, University of Arkansas
2010 – Dr. Chad P. J. Bennington, University of British Columbia
2009 – Dr. Victor A. Atiemo-Obeng, Dow Chemical
2008 – Dr. Fernando J. Muzzio, Rutgers University
2007 – Dr. Ramesh R. Hemraijani, Mixing & Reactor Solutions
2006 – Dr. Piero Armenante, New Jersey Institute of Technology
2005 – Dr. David S. Dickey, MixTech, Dayton OH
2004 – Dr. Suzanne M. Kresta, University of Alberta, Edmonton
2003 – Dr. Philippe Tanguy, Ecole Polytechnique, Montreal

I am very honored join this list of past recipients who have contributed so much to our understanding of mixing technology and whose insights I have been able to apply to processes.

Thank you to NAMF and all who made this possible. And thanks to all of you for listening.

2023 AIChE Annual Meeting – Orlando, FL – Paper 90B

86

This talk will be in the online proceedings, with all links live.

I will give a similar talk to the Process Development Division virtually on March 14, 2024

"Process Always Comes First" when successfully applying mixing technology

A personal reflection on 40 years
in industrial mixing and process development

Aaron Sarafinas
Sarafinas Process & Mixing Consulting LLC

2023 AIChE Annual Meeting – Orlando, FL – Paper 90B

Sarafinas Process & Mixing Consulting LLC

Lab

Plant

Bridging the gap

http://sarafinasprocess.com

aaron@sarafinasprocess.com

References

- ("AIM 2016") Kresta, S.M., A.W. Etchels III, D.S. Dickey, V.A. Atiemo-Obeng, **Advances in Industrial Mixing: A Companion to the Handbook of Industrial Mixing**, John Wiley and Sons, 2016.
- Berrier, J., D. Kolesar, A. Sarafinas, C. Teich, "Using DynoChem® to reduce batch cycle time in a gas-liquid reactor and commission a new unit," DynoChem User Group Meeting (2007).
- Bhatia, K., "Learnings from Personal Experiences in Process R+D", Presentation at Rohm and Haas, Spring House, June 2002.
- Bortz, M., K. Dadhe, S. Engell, V. Gepert, N. Kockmann, R. Muller-Pfefferkorn, T. Schindler, and L. Urbas, "AI in Process Industries – Current Status and Future Prospects," *Chimie Ingenieur Technik* **2023** 95(7), 975-988.
- Bourne J.R., "Mixing and the Selectivity of Chemical Reactions", **Org. Process Res. Des. Dev.**; 2003; 7(4) pp. 471-508.
- Bourne, J.R. and C.P. Hilber, "The productivity of micromixing-controlled reactions: effect of feed distribution in stirred tanks," **Trans IChemE** Vol. 68, Part A, January 1990.
- Brown, D.A.R, "Predicting the Critical Feed Time for Mixing Sensitive Reactions at Large Scales" NAMF MIXING XXVII June 2022.
- Chiang, L.H., B. Braun, Z. Wang, I. Castillo, "Towards artificial intelligence at scale in the chemical industry," *AIChE J* **68**(6), June 2022.
- ("HIM 2004") Paul, E.L., V.A. Atiemo-Obeng, S.M. Kresta, **Handbook of Industrial Mixing: Science and Practice**, John Wiley and Sons, 2004.
- Heise, C., A. Reeder, B. Lenich, B. Kane, D. Gibbs, G. Moody, K. Pleitt, L. Fernandez-Cerezo, M. Schofield, A. Johnston, *Chimica Oggi* 41(5) "CONTINUOUS BIOPROCESSING - THE JOURNEY HAS STARTED AND THE DESTINATION IS CLEARER THAN EVER" (2023).

References

- Hunt, K. and A. Sarafinas, “Virtuous not just Virtual Teams: Networking for Success at Rohm and Haas,” AIChE 2002 Spring National Meeting Paper 23g, New Orleans, LA, 2002.
- Janbon, S.L.M., A.R. Parsons, E. Gavi, G.K. Reynolds, “Effects of Scale, Equipment, and Operations on Agglomeration during a Reactive Crystallization”, **Org. Process Res. Des. Dev.**; 2019; 23(3) pp. 302-308.
- Jordan, C., C. Marton, Y. Agustinus, C. Wilbert, J. Perona, K. Ragazzo, and B. Holloway, “Scaling Down a Reactive Crystallization using Various Mixing Models,” **AIChE 2020 Annual Meeting**, Virtual, November 2020.
- Koestler, Don, “Life is a Batch” Presentation to AIChE Process Development Division (2001).
- Leng, DE, “Succeed at Scale Up”, **Chem Eng Prog**, pp. 23-31 (June 1991).
- Leng, D.E. and G.J. Quarderer (1982) “DROP DISPERSION IN SUSPENSION POLYMERIZATION,” **Chemical Engineering Communications**, 14:3-6, 177-201.
- Lerou, JJ and KM Ng, “Chemical Reaction Engineering: A Multiscale Approach to a Multiobjective Task”, **Chem Eng Sci**, 51(10), pp.1595-1614 (1996)
- O’Grady, D., M. Barrett, E. Casey, and B. Glennon, “The Effect of Mixing on the Metastable Zone Width and Nucleation Kinetics in the Anti-Solvent Crystallization of Benzoic Acid,” Trans IChemE Part A Che Res Des July 2007 85(A7) pp. 945-952.
- Oldshue, JY, **Fluid Mixing Technology**, McGraw-Hill, New York (1983).
- Oldshue, JY, DO Mechler and DW Grinnell (1982) “Fluid Mixing Variables in Suspension and Emulsion Polymerization” **Chem Eng Prog,** May 1982, 68-74.

References

- Paul, E.L., "Notes on Process Development: Pharmaceutical Intermediates and APIs," **AIChE Process Development Symposium**, Poconos-Lake Harmony, PA, 2003
- Plutschacket, M.B., Pieber, B., K. Gilmore, P.H. Seeberger, "The Hitchhiker's Guide to Flow Chemistry" *Chemical Reviews*, 2017, 117, 11796-11893.
- Sarafinas, A., "Applied Mixing Technology for Process Development: Scaling Down to Scale Up," course for Scientific Update, September 2023. (Course notes and addenda)
- Sarafinas, A., "Scaling Down to Scale Up: Advances in Using the Bourne Protocol and Optimization in the Micromixing-Mesomixing Space," **AIChE 2021 Annual Meeting**, Virtual, November 2021.
- Sarafinas, A., "Efficient Process Development in the Micromixing-Mesomixing Space: the Bourne Protocol and Beyond," **AIChE 2020 Annual Meeting**, Virtual, November 2020.
- Sarafinas, A., "Efficiently Characterize Process Scalability in the Micromixing-Mesomixing Space Using the Bourne Protocol," **AIChE 2019 Annual Meeting,** Orlando, FL, November 2019.
- Sarafinas, A. "Successful Process Development: Profitable Specialty Chemical Products Using Winning Work Practices," **AIChE Process Development Symposium**, Pocono Mountain, PA, June 2003.
- Teich, C.I. "Delivering Product When It's Needed, Where It's Needed," *Chemical Engineering Progress*, September 2023.
- Torbacke, M., A.C. Rasmusson, "Mesomixing in Semi-Batch Reaction Crystallization and Influence of Reactor Size" AIChE Journal 50 (2004) 3107-3119.
- Yamamoto, J.H., W.S. Park, A. Sarafinas, M.M. Cook, S.G. Maroldo, "Method of making lithium borohydride," US patent 7,029,641 (Apr. 18, 2006).

301a Experimental Study of Blending in Unbaffled and Baffled Tanks

Arthur W. Etchells III, Robert Hesketh, Austin Colon, Samual Goldberg, Nathaniel Nelson

The standard vessel for mixing in the process industry is a baffled tank. This design is very common especially for low viscosity fluids nonetheless, unbaffled vessels are found in industry and the laboratory. The reasons given for using unbaffled vessels are primarily concerns of cleaning the vessel between batches and the possibility of stagnation zones with high viscosity fluids particularly when properties change over the processing time.

In this presentation we will examine the effect of baffles on the blend time using one vessel; with and without baffles. The blend time was measured using the iodine color change method using a range of viscosities and Reynolds Numbers. The depth of the vortex was also measured. The same axial flow down pumping impeller is used. We identified two blend times in using unbaffled tanks, a short blend time for the outer zone and a much longer time for the inner vortex centered zone which defines the overall blend time for the vessel.

In this work we observed that the flow patterns are dramatically different. At high Reynolds Numbers with the baffled configuration we see the familiar turbulent diffusion throughout the vessel. Without the baffles the fluid segregates into two zones: a swirling inner zone near the shaft and a more turbulent outer zone. Videos of the color change experiments will be shown in the presentation.

The overall blend time at low Reynolds numbers is longer in the baffled tank at the same rotational speed compared to un-baffled tank and the difference decreases with Reynolds Number. At these low Reynolds Number the baffled blending is in the transitional flow regime between turbulent and laminar with a combination of eddies and striations. The last region of the tank to become completely mixed is located behind the baffles. At high Reynold's numbers the unbaffled blend times are longer than the baffled tank. The baffled tank blend times remain constant with Reynold's number, but the unbaffled blend times increase with Reynold's number. In this presentation we also report data for vortex depth as a function of Reynold's number.

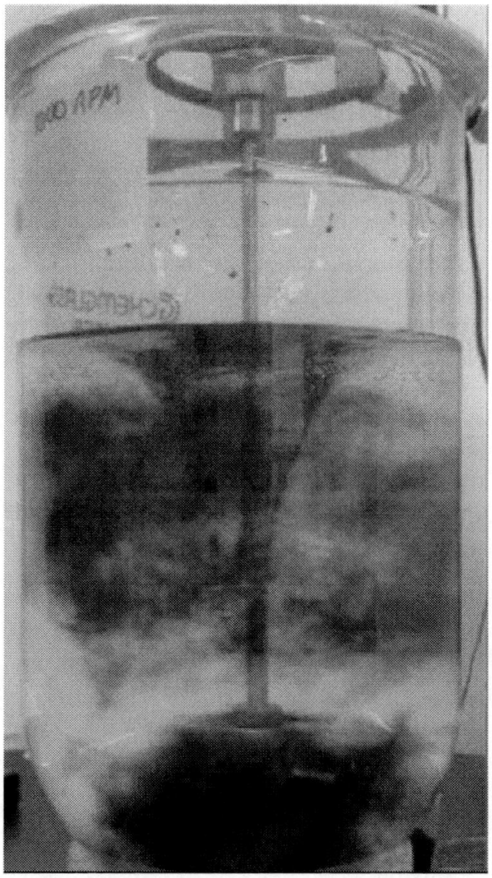

301b Empirical Study of Fluid Dynamics in Large Scale Fermentations Via Flow-Following Sensor Technology

Venkatesh Jayaraman

Biomanufacturing has shown the potential to replace or enhance products in a diverse set of traditional industries in the food, fuels, chemical and materials sectors with sustainable alternatives. Scaling up bioreactors promises economy of scale by amortizing CAPEX and OPEX over greater production volumes, and the need to deploy sustainable, scalable and innovative biomanufacturing solutions is becoming increasingly important. **Cargill** has a wide diverse line of **fermentation** products that are constantly expanding to meet demands in these sectors.

Remaining competitive in large-scale fermentation relies on the continuous development of innovative technologies to better monitor, understand and enhance the performance of our microorganisms in their highly complex bioreactor environments. Such developments reside with R&D as technology providers of our fermentation processes, leveraging deep scientific knowledge and pilot scale models. Despite the development of computational capacity, CFD simulations of large-scale aerobic fermentations remain a huge challenge due to the complex phenomena occurring at this scale. Fluid properties can change significantly during the fermentation, sometimes changing from Newtonian to non-Newtonian fluid. Many industrial bioreactors use state-of-the-art fixed sensors to measure physiochemical parameters, like pH, dissolved oxygen, conductivity, temperature, pressure etc., but acquisition of **spatially distributed parameters is still very complex**.

We have leveraged the flow-following sensors in Cargill pilot and industrial scale bioreactors to gather unprecedented empirical data on hydrodynamic behavior and gradients in pH, dissolved oxygen and temperature across the whole bioreactor. The measurement frequencies of these flow-following sensors are in the range of microseconds, which enable tracking fluid flow by analyzing the positional data along with the accurate representation of homogeneity of key parameters. Using this data, we have been able to measure hydrodynamic parameters like mixing time, circulation time and axial-velocities with spatial and temporal resolution in aerobic fermentation systems where the matrix is constantly changing. This has facilitated the development of data driven hydrodynamic compartment models that require much less computing power than CFD and can explain the mixing behavior inside pilot and industrial scale aerobic reactors.

Application of such emerging technologies has allowed Cargill to understand and draw conclusions on the impact of reactor design elements on the hydrodynamic characteristics and physiochemical gradients from a novel and alternative approach. These empirical models have the potential to improve mixing and aeration designs that will not only satisfy the oxygen transfer requirements but will also minimize energy input, identify & prevent dead zones, and reduce mixing time. All of these are major drivers in commodity fermentation economics.

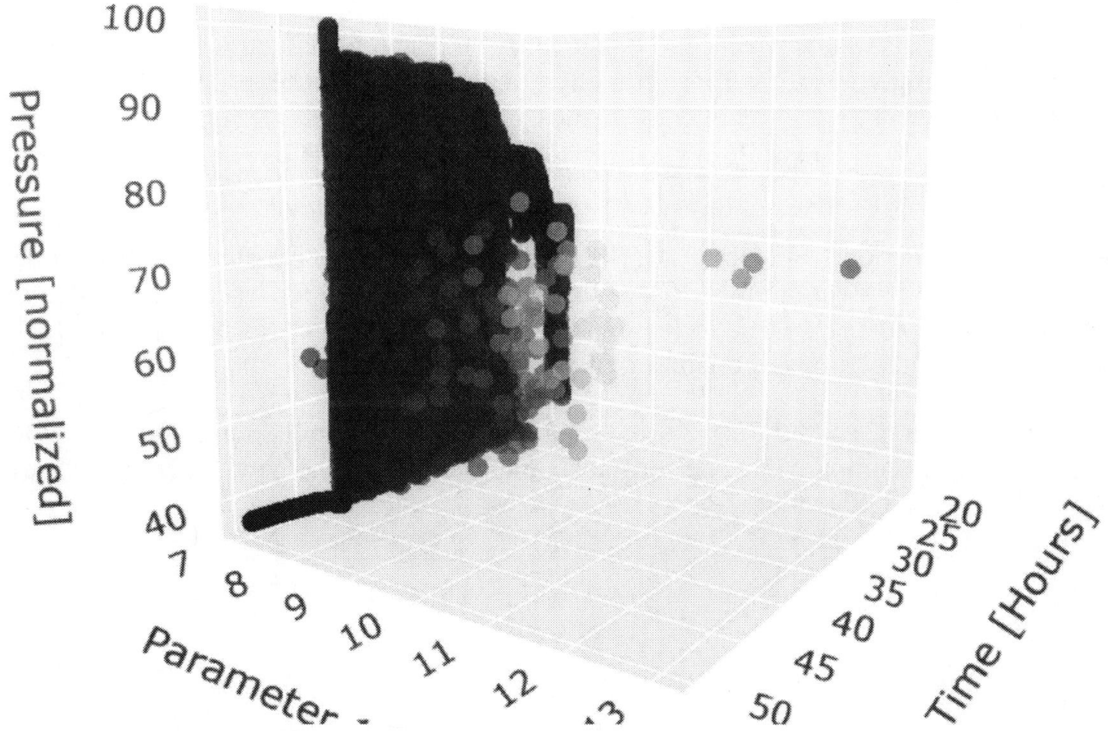

Pressure [normalized]
100
90
80
70
60
50
40
7
8
9
10
11
12
Parameter
Time [Hours]
20
25
30
35
40
45
50

301c Blending Performance of Hydrofoil Impellers with Tiplets

Richard Grenville, Jason Giacomelli, Ben Boyer

Grenville et al. (AIChE Annual Meeting, Phoenix AZ, 2022) presented data on the flow efficiency and trialing vortex properties of hydrofoil impellers with and without fins (tiplets) at the tips of their blades. The power drawn by the impellers was measured using an air-bearing and level arm torque meter while the mean flow and trailing vortex velocity and length scales were measured with a stereo particle image velocimeter.

Also, large eddy simulation CFD models of the three impellers were developed to compare the measured values of the Power number, Flow number and vortex properties and these also made predictions of the blend times. The conclusions drawn were that there was no appreciable difference in the performance of the three impellers tested on any basis.

Blend times have been measured for the three hydrofoils and the results presented here will provide another opportunity to compare their performance and to compare the new experimental results with the previously reported CFD values.

301d Comparison of Blending and Solids Sweeping Performance of Jet Mixers and Side-Entry Mixers in Vessels with Low Liquid Height to Tank Diameter Ratios

Aaron Strand, Jason Giacomelli

Vessels with low liquid height to tank diameter aspect ratios (H/T) are prevalent in the Energy sector and appear elsewhere across industries. Performing mixing operations in large vessels (i.e., greater than 10 m) with low H/Ts can be challenging and requires specialized equipment. Top-entry agitators are not economically viable. Two viable options include jet mixers and side-entry mixers. As a manufacturer of top-entry agitators, side-entry agitators, and centrifugal pumps, SPX FLOW has a vested interest in understanding the strengths and weaknesses of jet mixing and side-entry mixing and recommending suitable mixing equipment to its clients. An experimental comparison of blending and solids sweeping performance was conducted to further enhance and prove out the understanding of jet mixing and side-entry mixing in vessels with low H/Ts.

The investigation was conducted across two scales (1.07 m and 2.74 m). Three liquid height to tank diameter aspect ratios were investigated between the two scales (0.2, 0.3, and 0.4 H/T). Vertically angled and horizontal mixer orientations were investigated across both scales and types of mixing equipment. The jet nozzle diameter to tank diameter ratios (D/Ts) ranged between 0.005 and 0.013. The impact of jet nozzle geometries, including eductors, were investigated. The side-entry impeller diameter to tank diameter ratios ranged between 0.036 and 0.056 D/T.

Water was used as the bulk fluid. Blending trials in the 1.07 m scale were conducted with 6M NaOH and 6M HCl additions using phenolphthalein as a visual indicator. Blending trials in the 2.74 m scale were conducted with 0.1% volume fraction additions of a saturated NaCl solution and monitored via conductivity measurements. Solids sweeping trials were conducted at both scales using a slow-settling resin representative of sludge in energy or environmental applications and measured with grids marked on the vessel floors. The 1.07 m scale trials were conducted at similar theoretical mixing power as measured mixing powers were on the order of 1 W, were unreliable, and fell within instrumentation errors. The 2.74 m scale trials were conducted at equivalent measured system powers of 0.125 W/kg for blending and 0.109 kW for solids sweeping.

Blending and solids sweeping results were compiled and compared. Mean blend times and mean swept areas were calculated with 95% confidence intervals. Experimental blend time results were compared to theoretical blend time predictions (R. Grenville and J. Tilton, 1991; R. Kehn, 2011).

Conclusions drawn from the 1.07 m scale were:

1. Jet mixers and side-entry mixers provided similar blending performance at similar theoretical mixing power.
2. Jet mixers and side-entry mixers provided similar solids sweeping performance at similar theoretical mixing power.
3. Blending and solids sweeping performance was independent of jet mixer nozzle geometries, including eductors.
4. Blending performance was independent of mixer orientation (i.e., horizontal or angled).
5. Solids sweeping performance was dependent on mixer orientation.

Conclusions drawn from the 2.74 m scale were:

1. At equivalent system power, jet mixer blending performance trailed side-entry mixer.
2. At equivalent system power, jet mixer solids sweeping performance trailed side-entry mixer.
3. Blending and solids sweeping performance was independent of jet mixer nozzle geometries, including eductors.
4. Blending performance was considered independent of mixer orientation except there were

indications that angled mixers may perform better as H/T aspect ratios increased.

References:

1991. Grenville and J. Tilton, 'A new approach to correlation of blend time data from turbulent jet mixed vessels,' 1991.

1992. Kehn, 'Comparing Top Entry Versus Side Entry Agitator Performance in Low Viscosity Blending,' The Canadian Journal of Chemical Engineering, vol. 89, 2011.

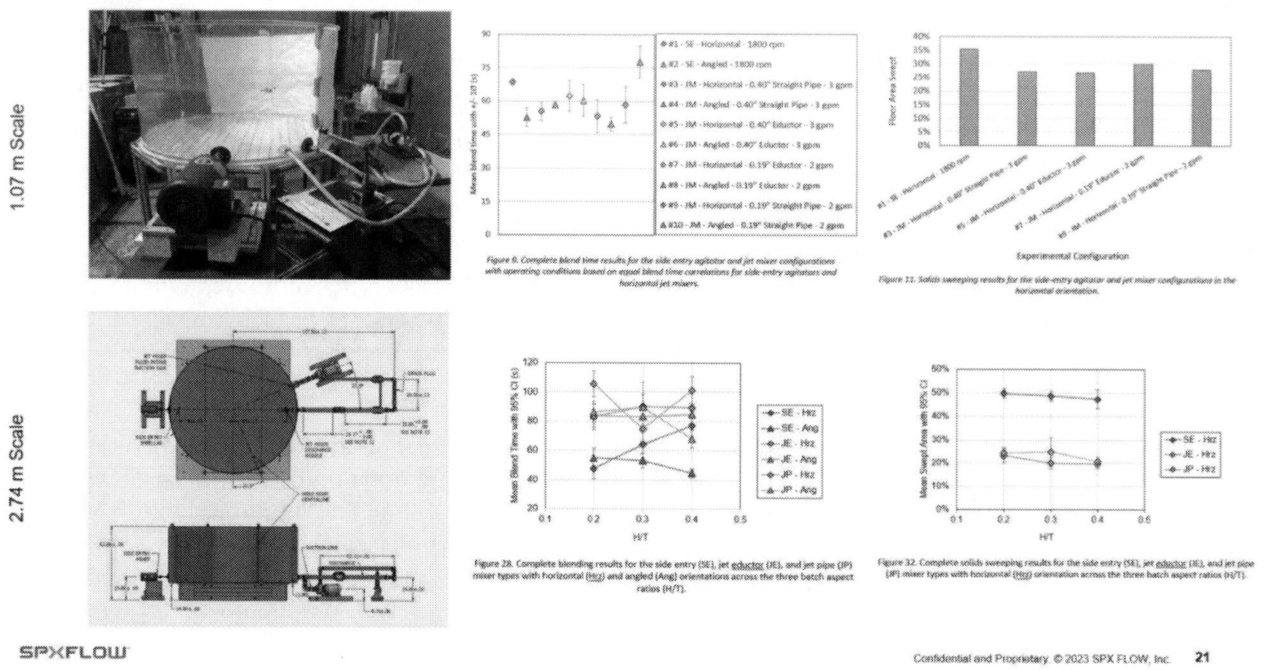

371a Direct Numerical Simulation of an Alternative Smx Element Orientation for Laminar Liquid-Liquid Mixing

Juan Valdes, Fuyue Liang, Lyes Kahouadji, Seungwon Shin, Jalel Chergui, Damir Juric, Omar Matar

Emulsions and liquid-liquid (L-L) mixing processes are of central importance in a broad range of industrial sectors, ranging from food and cosmetics to polymer and mineral processing, biotechnology, and pharmaceuticals (Valdes et al., 2022). Despite the fact that conventional high-energy methods, which involve the use of stirred vessels, remain the most common choice for liquid-liquid mixing, static mixers have proven to be a viable alternative; this is because they can retain, or even improve, the performance obtained through traditional methods at a lower capital and operational cost (Thakur et al., 2003). In particular, the SMX family of mixers has been successfully implemented in numerous industrial applications, handling highly viscous fluids and substantially large viscosity ratios (rvis ~ O(10^6)) (Hirschberg et al., 2009). Many former studies have explored the impact of varying the SMX's geometrical features in pursuit of an optimal or enhanced configuration (Liu et al., 2006; Singh et al., 2009; Heniche et al., 2005; Hirschberg et al., 2009; Meijer et al., 2012).

Nonetheless, performance metrics targeted in such studies mostly focus around pressure drop/energy consumption vs. miscible mixing criteria (e.g., concentration profiles and interfacial stretch) and flow streamline analysis, excluding relevant metrics for multiphase systems (e.g., droplet size distribution, DSD). In this sense, proposed modifications for the SMX (e.g., SMX+ design with reduced bar width) have essentially been conceived and tested for miscible mixing applications and are intrinsically expected to improve two-phase mixing performance as well. Hirschberg et al. (2009) explicitly tested a L-L dispersion scenario and verified that Sauter mean diameter correlations for the SMX are also applicable for the SMX+, but no further research followed on the dispersion performance. Furthermore, additional effects relevant in L-L mixing, such as the presence of surfactants, have been seldom included in performance studies given the extra layers of complexity that must be handled by the numerical framework. Despite this, the role of surfactants is of critical interest when dealing with L-L systems since the properties and stability of most structured goods (i.e., emulsion-based products) are heavily correlated with the manipulation of the interfacial tension between phases (Wong et al., 2015; Leal-Calderon et al., 2007; Valdes et al., 2022). In addition, previous studies have pointed out the deficiencies behind assuming a constant equilibrium interfacial tension and disregarding key interfacial and bulk mass transfer dynamics for the surfactant species (Haas, 1987; Lobry et al., 2011; Das et al., 2013). Based on the above, we consider pertinent to carry out a comprehensive mixing performance analysis with two-phase dispersion metrics as target objectives, where extra complexities (i.e., surfactants) are included in the computations to provide a more realistic, and industrially-relevant, depiction of the mixing operation.

We have utilized a high-fidelity, 3D direct numerical simulation (DNS) framework coupled with a state-of-the-art interface capturing algorithm based on a hybrid front-tracking level-set methodology, mounted on a massively parallelized computer architecture (Shin et al., 2017, 2018). Our code solves explicitly the unsteady dynamics of the free-interface, providing a wealth of information on the interfacial dynamics which are inaccessible via experiments or volume-averaged numerical approaches. In addition, our numerical framework resolves surfactant transport for both the interface and bulk phase through a set of convection-diffusion equations, and couples it to the interfacial tension via a Langmuir equation of state, as described thoroughly by Shin et al. (2018).

An alternative orientation (90° rotation) for the SMX element pair was tested (see Fig. 1), inspired by the spacer sections between elements trialed for similar static mixers (Abdolkarimi and Ganji, 2014). This rather simple modification is based on the premise that a considerable amount of daughter droplets and satellite structures are generated via hydrodynamically-driven breakage and capillary instabilities (Valdes et al., 2023). Even though geometrically-induced deformation strongly drives the formation of unstable ligaments and structures in the SMX, hydrodynamically-driven breakups frequently spawn numerous daughter droplets from a single unstable ligament, thus heavily contributing to the number of small and medium-sized droplets. These events are mostly independent of the flow field imposed by the SMX element (Valdes et al., 2023). In this study, the alternative SMX arrangement is contrasted against the dispersion performance of a standard 2-element SMX mixer, for both clean and surfactant-laden scenarios, varying relevant physicochemical parameters (elasticity and adsorption-desorption kinetics).

As expected by design, a lower pressure drop (thus lower power consumption) is obtained for the alternative configuration with a substantially different longitudinal velocity profile (due to the fewer crossbars in the second element). Results for the clean case with a multi-drop inlet set-up (see Valdes et al. (2023)) show that both configurations produce a comparable DSD but for a significantly lower pressure drop with the alternative mixer design. The latter is also associated with a slightly skewed shape towards larger sizes for the alternative SMX due to elongated but unbroken ligaments in the spacer section. In contrast, surfactant-laden simulations for a 3-drop inlet show a massive disparity in performance between the standard and alternative mixer configurations in terms of the DSD and number of drops when modelling soluble, highly desorptive surfactants. In such scenarios, the alternative SMX generates substantially wider distributions, skewed towards larger sizes, and a droplet count close to the clean case at high desorptive coefficients. Interestingly, poorly desorptive surfactants exhibit the least DSD variability for the alternative SMX, unlike the standard SMX where the least variable DSDs are achieved by the highly desorptive surfactants. Such a disparity is significantly reduced when assuming fully-insoluble surfactants, particularly at higher surfactant elasticities, where the droplet count and DSDs are seen to agree to a much higher degree.

Acknowledgements

This work is supported by the Engineering and Physical Sciences Research Council, United Kingdom, through the EPSRC PREMIERE (EP/T000414/1) Programme Grant. This work is also supported by the Colombian Ministry of Science, Technology and Innovation MINCIENCIAS, through a doctoral studentship for JV. O.K.M. acknowledges the Royal Academy of Engineering for a Research Chair in Multiphase Fluid Dynamics. We also acknowledge the HPC facilities provided by the Research Computing Service (RCS) of Imperial College London. D.J. and J.C. acknowledge support through computing time at the Institut du Developpement et des Ressources en Informatique Scientifique (IDRIS) of the Centre National de la Recherche Scientifique (CNRS), coordinated by GENCI (Grand Equipement National de Calcul Intensif) Grant 2022 A0122B06721.

References

J. P. ValdÅLes, L. Kahouadji, and O. K. Matar, "Current advances in liquid–liquid mixing in static mixers: A review," Chemical Engineering Research and Design, vol. 177, 1 2022.

R. K. Thakur, C. Vial, K. D. Nigam, E. B. Nauman, and G. Djelveh, "Static mixers in the process industries - a review," Chemical Engineering Research and Design, vol. 81, no. 7, pp. 787–826, 2003.

S. Hirschberg, R. Koubek, F.Moser, and J. SchÅNock, "An improvement of the Sulzer SMX™ static mixer significantly reducing the pressure drop," Chemical Engineering Research and Design, 2009.

S. Liu, A. N. Hrymak, and P. E. Wood, "Design modifications to SMX static mixer for improving mixing," AIChE Journal, 2006.

M. K. Singh, P. D. Anderson, and H. E. Meijer, "Understanding and optimizing the SMX static mixer," Macromolecular Rapid Communications, 2009.

M. Heniche, P. A. Tanguy, M. F. Reeder, and J. B. Fasano, "Numerical investigation of blade shape in static mixing," AIChE Journal, vol. 51, no. 1, 1 2005.

H. E. Meijer, M. K. Singh, and P. D. Anderson, "On the performance of static mixers: A quantitative comparison," Progress in Polymer Science, vol. 37, pp. 1333–1349, 2012.

S. F. Wong, J. S. Lim, and S. S. Dol, "Crude oil emulsion: A review on formation, classification and stability of water-in-oil emulsions," Journal of Petroleum Science and Engineering, vol. 135, pp. 498–504, 2015.

F. Leal-Calderon, J. Bibette, and V. Schmitt, Emulsion science: Basic principles, 2nd ed. Springer, 2007.

P. A. Haas, "Turbulent dispersion of aqueous drops in organic liquids," AIChE Journal, 1987.

E. Lobry, F. Theron, C. Gourdon, N. Le Sauze, C. Xuereb, and T. Lasuye, "Turbulent liquid-liquid dispersion in SMV static mixer at high dispersed phase concentration," Chemical Engineering Science, 2011.

M. D. Das, A. N. Hrymak, and M. H. Baird, "Laminar liquid-liquid dispersion in the SMX static mixer," Chemical Engineering Science, 2013.

S. Shin, J. Chergui, and D. Juric, "A solver for massively parallel direct numerical simulation of three-dimensional multiphase flows," Journal of Mechanical Science and Technology, vol. 31, no. 4, 4 2017.

S. Shin, J. Chergui, D. Juric, L. Kahouadji, O. K. Matar, and R. V. Craster, "A hybrid interface tracking – level set technique for multiphase flow with soluble surfactant," Journal of Computational Physics, vol. 359, pp. 409–435, 4 2018.

V. Abdolkarimi and H. Ganji, "CFD modeling of two immiscible fluids mixing in a commercial scale static mixer," Brazilian Journal of Chemical Engineering, vol. 31, no. 4, 12 2014.

J. P. Valdes, L. Kahouadji, F. Liang, S. Shin, J. Chergui, D. Juric, and O. K. Matar, "Direct numerical simulations of liquid–liquid dispersions in a SMX mixer under different inlet conditions," Chemical Engineering Journal, vol. 462, p. 142248, 4 2023.

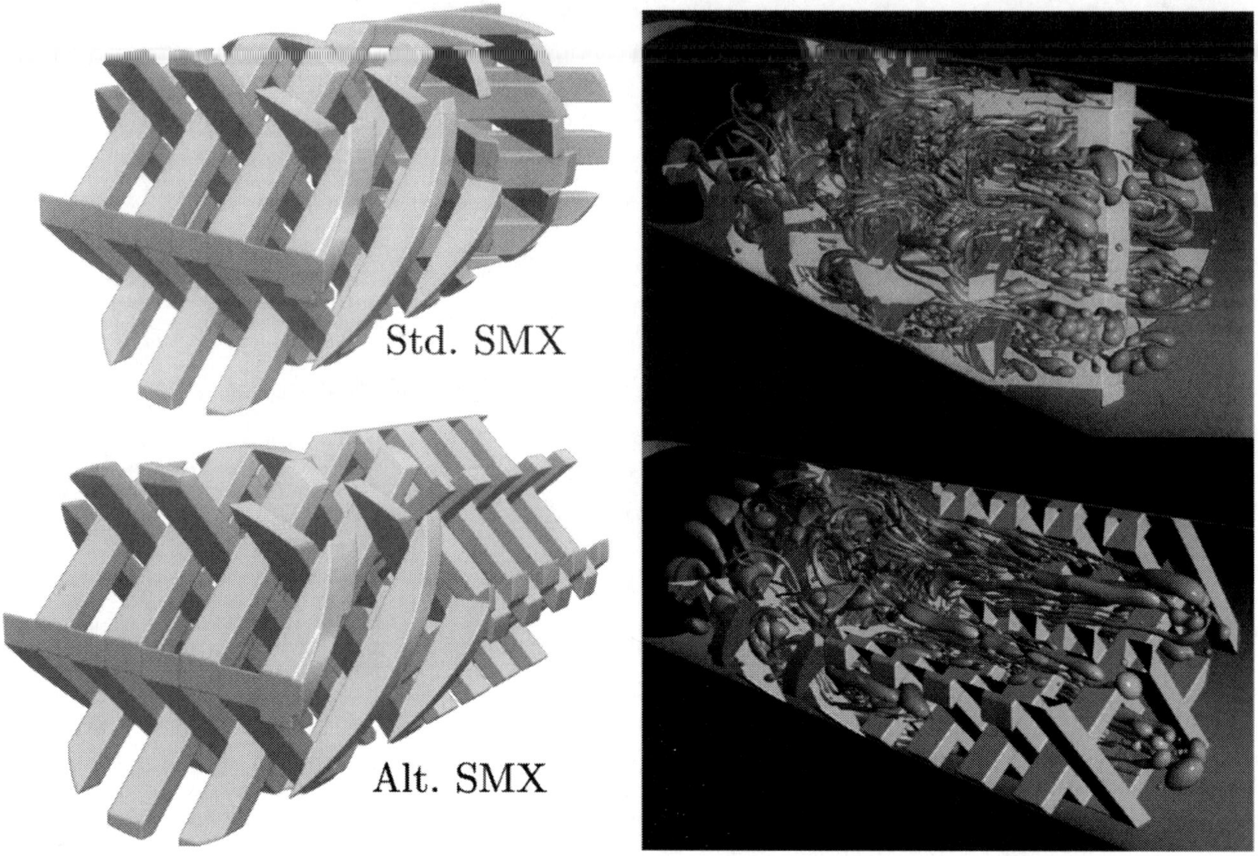

Std. SMX
Alt. SMX

371b Liquid-Liquid Dispersions: Modeling across Length Scales Using Eulerian-Lagrangian Conversion

Brian DeVincentis, John Thomas, Christopher Tyler

Liquid-liquid dispersions are abundant across many industries; these include food emulsions, cosmetic creams, pharmaceutical products, and many more. Many systems that produce these dispersions receive two separate immiscible liquids as input, then use mechanical energy to break up one phase into small droplets. Modeling these systems using CFD can aid in optimization, but the large range of scales involved is challenging to model. The system can initially be modeled with an Eulerian approach as the interface between the immiscible phases is on the macroscopic scale. At the end of the process, Lagrangian particles best capture the state of the system since the dispersed phase has been broken up into microscopic droplets. During the process, the system will have both macroscopic and microscopic regions meaning the model must be able to represent both and convert between them.

Here we present and validate an algorithm for converting the dispersed phase between Eulerian and Lagrangian representations. Macroscopic regions are converted to particles as the dispersed phase is broken up. Meanwhile, particles are converted to macroscopic regions as they coalesce. First, we apply the model to turbulent jet breakup, a well-characterized system, comparing against published data. Next, it is validated against static mixers and agitated tanks, comparing to experimental results. Finally, performance is analyzed, demonstrating the acceleration gained by using GPU hardware.

371c Machine Learning for Prediction and Optimization in Surfactant-Laden Liquid Dispersions

Fuyue Liang, Juan Valdes, Sibo Cheng, Lyes Kahouadji, Seungwon Shin, Damir Juric, Jalel Chergui, Omar Matar

Drop size and its distribution are of central importance to a broad range of industrial applications where liquid-liquid dispersion is heavily involved: from food and cosmetics to drug development and biotechnology. Numerous studies have been dedicated to the establishment of predictive correlations regarding (mean and maximum) drop size by considering various flow conditions and physical properties of the liquid phases [Calabrese et al., 1986, Zhou and Kresta, 1998a, Kraume et al., 2004, Zainal Abidin et al., 2015]. Moreover, studies have investigated the effect of design of commonly-used mixers on the drop size, covering various types of agitated mixers [Zhou and Kresta, 1998b, Pacek et al., 1999, Giapos et al., 2005, Sechremeli et al., 2006, Ghotli et al., 2019]. The case of static mixers, however, has enjoyed comparatively far less attention in the literature. In any case, it is challenging to exhaustively explore the effect of physical properties and mixer design on the drop size given the sensitivity of the latter to a slight change in these factors.

In addition, the prevalent scenario in industry, wherein surface active agents (surfactants) are present as contaminants or additives, elevates the complexity of predicting the surfactant-covered drop size (distribution). It is unsurprising, therefore, that no correlation thus far has been established to predict the drop size in the presence of surfactant. Fortunately, the rapid development of Artificial Intelligence (AI), data science, and machine learning provides a powerful toolkit to address the complexities described above in the field of multi-phase flow [Zhu et al., 2022]. Inspired by prior work, machine learning techniques are applied herein for the development of predictive models of the drop size given either the surfactant properties or mixer configurations.

In this study, high-fidelity simulations of surfactant-laden liquid dispersions obtained via use of static and agitated mixers are carried out to thoroughly investigate the underlying physical mechanisms governing the mixing dynamics (Figure 1). Our state-of-the-art numerical framework comprises a robust hybrid interface-tracking algorithm, along with a well-validated solver for surfactant transport in multi-phase flow field [Shin et al., 2018]. This grants us access to an unprecedented level of detail of the interfacial evolution in the presence of surfactants. Taking advantage of the drop size tracking and corresponding interfacial surfactant concentration profiles, the present work seeks to exploit supervised learning methods (i.e., Random Forest, XGBoost) to explore how the drop size is modulated dynamically by the interfacial tension, which is dependent on the concentration and physicochemical nature of the surfactant present. Our results show that, for both mixers, the drop size is strongly correlated to the average interfacial surfactant concentration while physicochemical properties and initial conditions (i.e., desorption coefficient, initial interfacial and bulk concentration) exert a relatively weak influence. Furthermore, semi-empirical drop size correlations from previous work in the literature are encoded into our machine learning algorithms with the aim of identifying the deviation between clean and surfactant-laden drop size predictions.

Finally, coarse simulations of mixers with various configurations are on-going to generate data to train a surrogate model for rapid prediction of mixer performance and thereby leading to mixer design optimization. More specifically, take agitated mixer as an example, effort is made on the pitched blade turbine, one of the most commonly used impeller type in industrial applications, where the effect of multiple design parameters (namely pitch angle, blade width, blade length, blade number, and clearance) will be studied. The surrogate model to be developed is expected to either predict drop size (or interfacial area) when certain mixer configuration is known, or provide guidance on mixer design given a specific metric, say, drop size distribution.

Acknowledgement

This work is supported by the Engineering and Physical Sciences Research Council, United Kingdom, through the EPSRC MEMPHIS (EP/K003976/1) and PREMIERE (EP/T000414/1) Programme Grants. O.K.M. acknowledges funding from PETRONAS and the Royal Academy of Engineering for a Research Chair in Multiphase Fluid Dynamics. We acknowledge HPC facilities provided by the Research Computing Service (RCS) of Imperial College London for the computing time. D.J. and J.C. acknowledge support through HPC/AI computing time at the Institut du Developpement et des Ressources en Informatique Scientifique (IDRIS) of the Centre National de la Recherche Scientifique (CNRS), coordinated by GENCI (Grand Equipement National de Calcul Intensif) Grant 2022 A0122B06721.

References

Richard V Calabrese, TPK Chang, and PT Dang. Drop breakup in turbulent stirred-tank contactors. part i: Effect of dispersed-phase viscosity. AIChE Journal, 32(4):657–666, 1986.

Reza Afshar Ghotli, Mohammad Reza Abbasi, AmirHossein Bagheri, Abdul Aziz Abdul Raman, Shaliza Ibrahim, and Huseyin Bostanci. Experimental and modeling evaluation of droplet size in immiscible liquid-liquid stirred vessel using various impeller designs. Journal of the Taiwan Institute of Chemical Engineers, 100:26–36, 2019.

A Giapos, Chrysostomos Pachatouridis, and Michael Stamatoudis. Effect of the number of impeller blades on the drop sizes in agitated dispersions. Chemical Engineering Research and Design, 83(12):1425–1430, 2005.

Matthias Kraume, Ansor G¨abler, and Kai Schulze. Influence of physical properties on drop size distribution of stirred liquid-liquid dispersions. Chemical Engineering & Technol- ogy: Industrial Chemistry-Plant Equipment-Process Engineering-Biotechnology, 27(3): 330–334, 2004.

AW Pacek, S Chamsart, AW Nienow, and A Bakker. The influence of impeller type on mean drop size and drop size distribution in an agitated vessel. Chemical Engineering Science, 54(19):4211–4222, 1999.

D Sechremeli, A Stampouli, and M Stamatoudis. Comparison of mean drop sizes and drop size distributions in agitated liquid–liquid dispersions produced by disk and open type impellers. Chemical Engineering Journal, 117(2):117–122, 2006.

Seungwon Shin, Jalel Chergui, Damir Juric, Lyes Kahouadji, Omar K Matar, and Richard V Craster. A hybrid interface tracking–level set technique for multiphase flow with soluble surfactant. Journal of Computational Physics, 359:409–435, 2018.

Mohd Izzudin Izzat Zainal Abidin, Abdul Aziz Abdul Raman, and Mohamad Iskandr Mohamad Nor. Mean drop size correlations and population balance models for liq- uid—liquid dispersion. AIChE Journal, 61(4):1129–1145, 2015.

Genwen Zhou and Suzanne M Kresta. Correlation of mean drop size and minimum drop size with the turbulence energy dissipation and the flow in an agitated tank. Chemical Engineering Science, 53(11):2063–2079, 1998a.

Genwen Zhou and Suzanne M Kresta. Evolution of drop size distribution in liquid–liquid dispersions for various impellers. Chemical Engineering Science, 53(11):2099–2113, 1998b.

Li-Tao Zhu, Xi-Zhong Chen, Bo Ouyang, Wei-Cheng Yan, He Lei, Zhe Chen, and Zheng- Hong Luo. Review of machine learning for hydrodynamics, transport, and reactions in multiphase flows and reactors. Industrial & Engineering Chemistry Research, 61(28): 9901–9949, 2022.

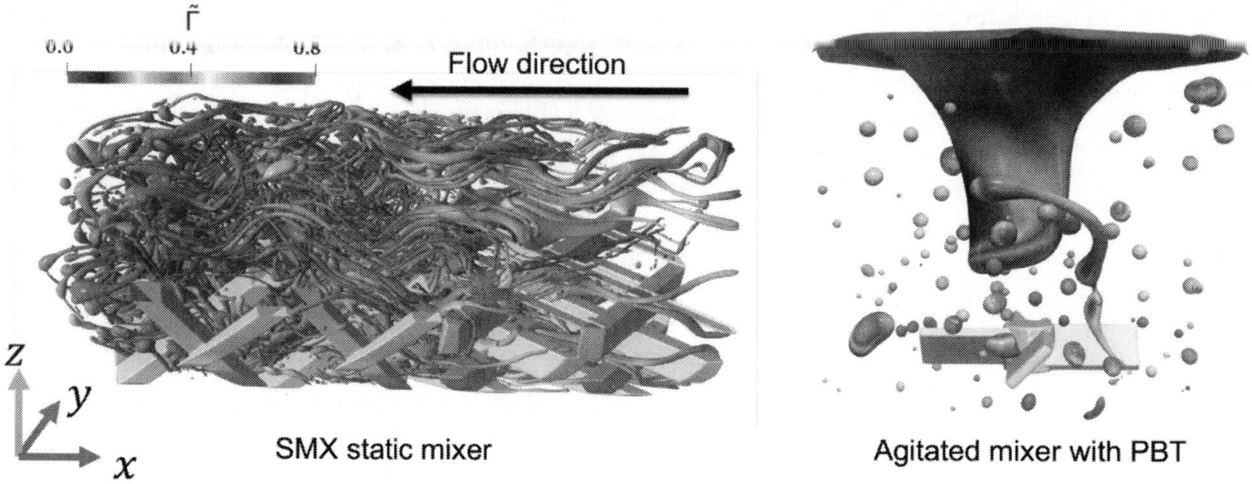
Γ̃
0.0
0.4
0.8
Flow direction
z
y
x
SMX static mixer
Agitated mixer with PBT

371d CFD Characterization of a Vertically Stacked 7-Stage Continuous Crystallizer Cascade

Jody Albertazzi, Ajinkya Pandit, Giovanni Aprile, Torsten Stelzer, Allan S. Myerson

Mixing plays a crucial role in crystallization processes to obtain e.g., a homogeneous supersaturation profile and a better control on the particle size distribution. Parameters such as the impeller type, size, position, and speed can heavily influence the mixing in Mixed Suspension Mixed Product Removal (MSMPR) crystallizer, the most widely adopted type among the continuous crystallizers. In the presented work, a novel stacked 7-tank MSMPR cascade is studied where the slurry is displaced vertically, by intervention of two periodically inflating diaphragms, in the first and last stage. It is of fundamental importance to understand and optimize the impact of the impeller configuration on the hydrodynamics inside the spherical tanks of the novel crystallizer aimed at minimizing the settling of solids and ensuring swift micromixing of antisolvent. Therefore, Computational Fluid Dynamics (CFD) was used to numerically study different impeller configurations (type, size, position) to map out the crystallizers mixing characteristic and identify the optimized setting (lowest impeller speed to minimize crystal breakage). Simulations were conducted utilizing Ansys Fluent software with an unstructured computational grid employing the RNG k-ε model. The results in this work represent the first CFD study for the novel crystallizer design. It also addresses the demand of the scientific community to investigate the hydrodynamics in crystallizers.

371f Scale-Down and Scale-up of a Slurry Mixing Process

Richard Grenville

Scale-up is a common activity in process development. The goal is to develop the understanding of how a particular unit operation affects the outcome of the process in small scale equipment and then design equipment for the full-scale plant based on this understanding.

If a piece of equipment in a plant is not operating as designed, the operation can be scaled down, recreating the problem at small scale to develop an understanding of its causes. Then further tests can be carried out to develop a solution to the problem at small scale followed by the design of equipment to implement at plant scale.

A case study will be presented which will discuss how mixing issues were created, due to changes in the operating procedures, in the production of a viscous, non-Newtonian, high solids concentration slurry. The changes resulted in the slurry being more viscous and shear-thinning. The agitator was not capable of moving the slurry in regions far from the impeller blades and severe stagnation and fouling was observed. The process would have to be shut down on a regular basis so that the vessel could be cleaned.

Scale-down studies were carried out in a combined stirred tank / viscometer developed for the project enabling a mixing test to be performed identifying the degree of agitator required to eliminate stagnant regions and, at the same time, to measure the slurries rheology. The results were scaled-up and the new agitator has been working as specified for nearly five years.

Finally, some of the considerations that must be taken into account when making rheology measurements, especially for high concentration slurries, will be addressed.

456a CFD Simulation of Impulse Mixers

Iman Mirzaee, Johannes Wutz, John Thomas, Masoud Ahmadi, Saeed Jafari Kang, Shahab Golshan, Fabrice Schlegel, Pablo A. Rolandi

Mixing is one of the most common unit operations in pharmaceutical industry. In recent years, pharmaceutical industry has shifted from using empirical techniques to numerical methods due to their cost saving and increased level of details in their analysis. A complete three-dimensional in silico analysis of mixing process aids the selection of working conditions and agitator designs. Although in silico modeling is more common in recent years, new and creative designs of mixers require more attention to the CFD models to ensure the validity of results.

In the current work, results of a CFD model using an imPULSE agitator in a mixing tank geometry is presented. The model is assessed in terms of numerical capabilities. The model then applied to a case study to fully replace experimental studies.

456b Scale up/Scale Down of Impulse Mixer Using CFD

Mustafa Kaddoura, Hossam Metwally, Kringan Saha, Chadakarn Sirasitthichoke, Robert Blizzard

imPULSE mixers provide unique flow and mixing characteristics that make them attractive for certain mixing applications especially at medium to larger scales. They also inherently are less susceptible to vortexing and air entrainment from the top. Instead of relying on the typically rotating impeller to provide the agitation, they rely on an oscillating mixer that can provide the required power by unit volume (P/V), as shown in Figure 1.

From a modeling standpoint, they do pose unique challenges though. The continuously changing geometry because of the mixer oscillation requires the implementation of a moving and deforming (MDM) technique that must be both efficient (i.e., allow for fast simulation) and maintain good accuracy for the eventual species mixing part of the simulation (i.e. avoid numerical diffusion). An additional complexity that is inherent to imPULSE mixers is the presence of "flaps" within the mixers that "open" during the upward stroke and "close" during the downward stroke. This on-off behavior of the flaps augments the effect of the mixer motion on the flow field within the vessel and enhances the mixing process.

In this current work, we will present a novel way to represent the presence of the flaps using a porous media approach. The porous media representing the presence of the flaps will have a time varying flow resistance that drops to zero during the upward stroke and has an artificially high value during the downward stroke. In addition, the "mesh layering" technique will be used to re-mesh the domain as the mixer oscillates. It is also noted that this method is the most efficient and accurate among re-meshing techniques and is ideal for oscillatory motion in general.

Finally, three different scales will be simulated and compared using the proposed approach, namely the current size of 5,000-L, 3,000-L, and a 250-L size with respect to their P/V and blend time.

Figure 1: imPULSE Mixer

Figure 2: CFD Mesh

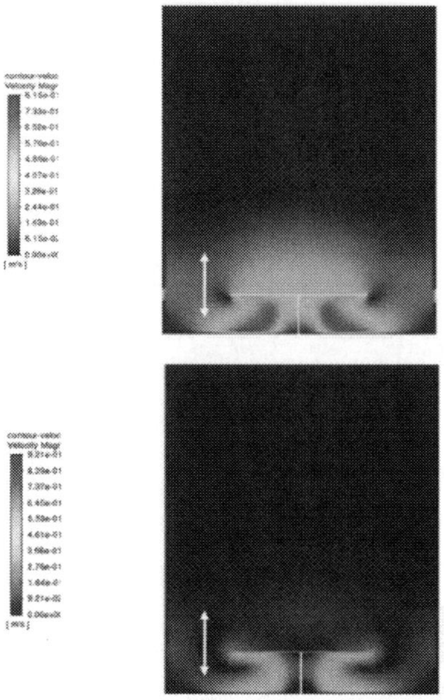

Figure 3: Velocity contours at two different time instants

456c Mixing Time and Fluid Flow Profiles in Asymmetric Flow Jet Mixing Reactors

Faiz Khan, David Mualen, Barbara E. Wyslouzil, Nicholas Brunelli, Jessica Winter

Microreactors/mixers offer promise for nanomanufacturing through scale-out versus scale-up options. Microreactors have been employed by the pharmaceutical industry for production of drug delivery vehicles, and most notably Covid mRNA vaccines. A common microreactor design involves impingement of jets with each other or with a crossflow stream. Many of these reactor designs necessitate equal jet flow rates to prevent asymmetries in the impingement zone that can lead to undesired flow patterns. However, asymmetric flow (unequal inlet flow rates) may be desired for a variety of reasons, such as use of delicate or expensive cargoes, to reduce toxicity and optimize safety of selected reactants, or to control supersaturation and solvent quality to maximize product quality.

Here, we examined mixing times and flow patterns in a jet mixing reactor under asymmetric flow conditions. The jet mixing reactor explored was originally developed for gas phase synthesis and demonstrated recently for synthesis of zeolites[1]. The reactor has a crossflow geometry, consisting of a main line onto which two opposing jets impinge. The resulting product line exits the reactor as a continuation of the main line. This reactor has recently been used for manufacturing of Ag@Au nanoparticles, polymer nanostructures, quantum dots, and copper nanoparticles, demonstrating its versatility. Mixing time was experimental evaluated using the Villermaux-Dushman reaction[2], which consists of competing reaction sets with fast and slow kinetics whose products can be characterized using UV-Visible absorbance spectroscopy. The ratio of these products can be correlated to the mixing time through known equations[3]. Flow in the reactor was observed using dye streams. These results were corroborated and expanded upon using COMSOL simulations of the flow field. Mixing time in simulations was estimated by evaluating the point at which concentration reached 80% of the expected equilibrium value.

Concordance between experimental and theoretical data sets was high. Under symmetric flow, mixing times ranged from ~ 5-40 ms depending on reactor geometry (Channel diameters: 0.01-0.04") and operational flow rate (5-32 ml/min). Under asymmetric flow, mixing times were increased relative to symmetric mixing conditions, and at ratios of 5:1 for the jet: main line flow rates homogenous kinetics for polymer nanoparticle precipitation could no longer be obtained and mixing time was insensitive to flow rate for total inlet flow rates > 35 ml/min. In more detailed examination of flow fields within these reactors, it was determined that mixing was enhanced by rotational vortices formed in the direction perpendicular to flow of the existing fluid downstream of the impingement zone. At high jet: main line flow ratios (i.e., ~ 5-10), back flow emerges into the main line that limits mixing. These data suggest that operation in asymmetric conditions is possible, but that there is an upper limit to operation without additional design improvements.

1. A. Parulkar and N. A. Brunelli, *Industrial and Engineering Chemistry Research*, 2017, **56**, 10384-10392.
2. J. Commenge and F. Laurent, *Chem. Eng. Process.*, 2011, **50**, 979-990.
3. E. Arian and W. Pauer, *Chem. Eng. Res. Des.*, 2021, **175**, 296-308.

456d The Importance of Selecting Appropriate Scale-up Concepts: One Example of Mixing Issues in an Industrial Fermentation Process

Sören Bernauer

Mixing characteristics of fermentation reactors play a crucial role in an efficient operation. This is particularly important when processes are scaled-up into production scale reactors, with dimensions strongly diverging from small- and pilot-scale and leading to larger mixing times.

In this work, a fungal fed-batch process was scaled-up into an 80 m^3 production-scale fermenter based on common scaling approaches. Against the expectations from the process design in small scale, the process showed a different growth behavior in large scale with a strong increase in biomass and a decrease in productivity. In the root cause analysis for this behavior, the liquid flow of the reactor was simulated by CFD coupled with a metabolic model to mimic the glucose consumption and resulting in a spatial concentration distribution in the reactor. The results revealed inhomogeneities of the glucose due to a suboptimal position of the substrate feed inlet. Based on these findings, a sensitivity of the organism towards glucose gradients was considered as possible reason for the observed behavior in growth. In order to confirm this hypothesis, the glucose inhomogeneities were scaled down by applying the substrate feed discontinuous as pulses in a small-scale reactor with ideal mixing behavior. This method allows to emulate the spatial concentration differences between high glucose concentrations around the feeding position and low concentration in less mixed zones, over time. The resulting process in small scale showed a similar behavior in terms of biomass growth and decrease in productivity as the unsuccessful process in large scale, confirming the hypothesis. Based on these findings, several measures were taken to reduce mixing inhomogeneities in large-scale. The following production campaign was successful with an increased productivity as anticipated from the small-scale.

This example illustrates the importance of including the appropriate scaling concepts for a process transfer. Some organisms and processes might be influenced by parameters that are changing during scale-up. Especially when scaling into large industrial scale reactors, intermediate scales are important to reveal potential sensitivities but are mostly not considered due to a lack of time and resources.

456e Methodology for Mixing Scale-up: Safety, Quality and Reliability

Ankur Kapil, Michael Hamlin

Poor mixing alone is worth 1-10 billion$ to the US chemical industry alone[1]. In this presentation we would discuss some of our past failures, where the batch didn't meet the "Right First Time" quality and process requirements. To overcome these issues, we have developed a global database across sites including detailed vessel dimensions, vessel CAD drawings. We have deployed the equipment database in a Web app using Django framework. This webapp will improve access to the equipment capacity, dimensions of the baffles, impellers, tanks, heat transfer etc. for site Engineers, technicians, and Chemical Development chemists. Traditional mixing scale-up is based on a single parameter (Tip speed, Power/Volume) for scale-up. However, this is highly dependent on the application and mixing time and lengths. Critical process parameters for mixing are entirely different for hydrogenation, crystallization, or emulsification. Also, it is very important to understand which mixing length and time scale (micro, meso, or macro) is the limiting factor. Bourne[2] protocol is utilized to narrow down the mixing timescale and length for the particular product & process. Three different software packages are utilized to improve our process understanding of mixing behaviour across scale: Dynochem®, Vismix®, & CFD (OpenFOAM®). These packages differ in terms of complexity, accuracy, & simulation Cost. We would discuss the advantages and limitations of each of the software in this presentation. The methodology we have developed utilizes the Critical process metrics and scale-down experiments to de-risk scale-up.

This methodology has been utilized successfully to scale-up multiple products "Right First Time". A case study for successful application of this methodology to Multiphase reaction (Hydrogenation) would be demonstrated in this work. The reaction was scaled-up successfully from lab scale Parr (300ml) reactor to 100-gallon scale and finally to 500-gallon reactor. Scale-down experiments were used to demonstrate control methodology and assumptions before scale-up. The total value of this product including multiple batches to the business was 3.35MM$ with the right quality specifications.

References:

1. Kresta, S. M. & Brodkey, R. S. Turbulence in Mixing Applications. in *Handbook of Industrial Mixing* 19–87 (John Wiley & Sons, Ltd, 2003). doi:10.1002/0471451452.ch2.
2. Bourne, J. R. Mixing and the Selectivity of Chemical Reactions. *Org. Process Res. Dev.* **7**, 471–508 (2003).

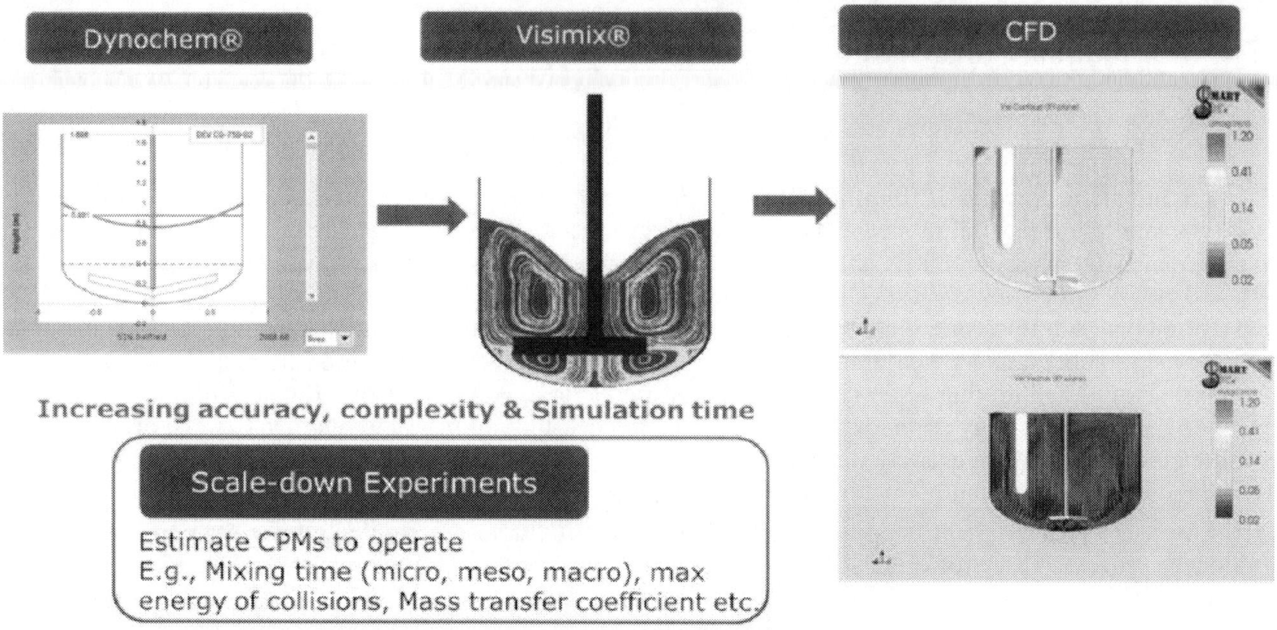
Dynochem®
Visimix®
CFD
Increasing accuracy, complexity & Simulation time
Scale-down Experiments
Estimate CPMs to operate
E.g., Mixing time (micro, meso, macro), max energy of collisions, Mass transfer coefficient etc.
1.20
0.41
0.14
0.05
0.02

423g High-Fidelity Simulation Validation for Cases with Phase-Change, Chemical Reaction, and Multiphysics

Nima Yazdanpanah

The Verification and Validation practice is one of the essential components of the modeling governance. Simulation of linear, simple, single-phase physical systems are usually "easy" and provide results with high level of accuracy. The level of the high-fidelity simulations, model setup, system and parameter definition/estimation, and validation becomes more complex and challenging for nonlinear system. Some examples of these systems are such as systems with variable viscosity (e.g. polymerization), phase change and slurry (e.g. crystallization), and multiphysics (e.g. multiphase hydrogenation reactions with heat and mass transfer).

In this presentation, we will demonstrate some industrial cases with varied complexities where modeling and simulation were used for sensitive or complex systems. Few validation cases will be discussed for stirred cooling crystallization, a complex multiphase reaction set in a large plug flow reactor with static mixer, a multiphase bioreactor and hydrogenation.

423b Comparing Predicted Just Suspended Speed Versus Experimental Just Suspended Speed for Axial Flow Impellers in a Stirred Vessel

Sarah Lanzafame

Mixing applications that require suspension of solids in a liquid are common in just about every major industry. In many cases, the mixer just needs to keep the solids from settling, in order to allow the slurry to be transported from tank to tank. This is known as the just suspended condition, or the Zwietering criterion, which states that all particles have some vertical motion and do not remain on the bottom of the vessel for more than 1 to 2 seconds. Zwietering studied suspension of particles in stirred tanks to develop a correlation to predict the just suspended speed for a given impeller style (equation 1). This correlation has been an industry standard for decades. One limiting factor for the correlation is the impeller "S" constant, which is specific to the impeller style, impeller diameter to tank diameter ratio (D/T) and the impeller's location relative to the tank bottom. Thus, a specific "S" value is required for each geometry. Grenville, Mak and Brown developed a new correlation that brings in the impeller geometry (off bottom distance ratio) and categorizes impellers in two categories; pitched blade turbines and hydrofoils, since these are the most commonly used impellers for suspension applications. This correlation (equation 2) has similar elements to the Zwietering correlation.

The experimental study presented at the 2021 AIChE Annual Conference contained a Lightnin A200 (pitched blade turbine) and a Lightnin A310 (hydrofoil) which were tested over three volume scales to experimentally determine the just suspended speed condition. Vertical tanks with four anti swirl wall baffles and a torispherical dished bottom were used for the study. The three scales were a 445 mm diameter vessel (90 liters), a 597 mm diameter vessel (220 liters) and an 800 mm diameter vessel (530 liters). A batch aspect ratio of 1.3 was used in each case with a single impeller near the bottom. The impellers were selected to match as closely as possible an existing geometry where an "S" value for the Zwietering correlation existed. Two types of glass beads were tested in each case: one with a particle size of 147 microns and one with a particle size of 352 microns all with a solids concentration tested at 1.0 wt%. The mixer shafts were connected to strain gauge torque cells to measure shaft torque and optical tachometers to measure shaft speed. Although not a focus of this paper, cloud height was also measured for each condition tested. The Zwietering criterion was used to determine the just suspended speed condition by first increasing the speed of the mixer to achieve this condition and then reducing speed subsequently to ensure the condition was not overshot.

At the last AIChE conference in 2022, the quantity of impeller diameters tested for the hydrofoils increased and ranged from 5.2" to 8.8" in diameter and varying concentration from 1.0 wt% to 15.0 wt% (in 5.0 wt% increments) for the smallest vessel scale.

Since the last presentation at the AIChE conference, the extended work to be conducted to further this study are completing the two larger vessel scales using the various impeller diameters scaled up using the same D/T ratio as well as varying the solids concentrations from 1.0 wt% to 15.0 wt% (in 5.0 wt% increments). A more detailed statistical analysis will be completed and presented once all the scales are tested. The expected result is to be consistent to what was seen thus far, GMB and Zwietering both predicting similar results to what was experimentally collected for the larger particle size. The observation from last time was that the smaller particle size was the most conservative (easier to suspend than predicted) for both correlations, and this will be examined at these larger scales to see if this trend is consistent.

$$(\boldsymbol{EQ-1}):\ N_{js} = S\frac{\nu^{0.10} d_p^{0.20} X^{0.13} \left(\frac{g\Delta\rho}{\rho_L}\right)^{0.45}}{D^{0.85}}$$

$$(\boldsymbol{EQ-2}):\ N_{js} = \frac{z}{N_p^{\ 1/3} D^{2/3}}\left[\frac{g(\Delta\rho)}{\rho_L}\right]^{0.5} X_v^{\ 0.154} d_p^{\ 1/6}\left(\frac{C}{D}\right)^{0.1}$$

Variable	Description	Units
N_{js}	Impeller Speed for Just-Suspended State of Particles	rev/sec
S	Dimensionless Impeller Constant	No units
v	Kinematic Viscosity	m^2/s
g	Gravitational Constant	9.8 m/s^2
d_p	Particle Size	m
X	Mass Ratio of Suspended Solids to Liquid x 100 (%)	(kg solids / kg liquid) * 100
Δρ	Density Difference – Solids Density minus Liquid Density (ρ_s - ρ_L)	kg/m^3
ρ_L	Density of Liquid Phase	kg/m^3
D	Impeller Diameter	m
z	Dimensionless Constant in GMB Equation	No units
N_p	Impeller Power Number	No units
X_v	Volume of Solids per Unit Volume of Slurry x 100 (%)	Percentage

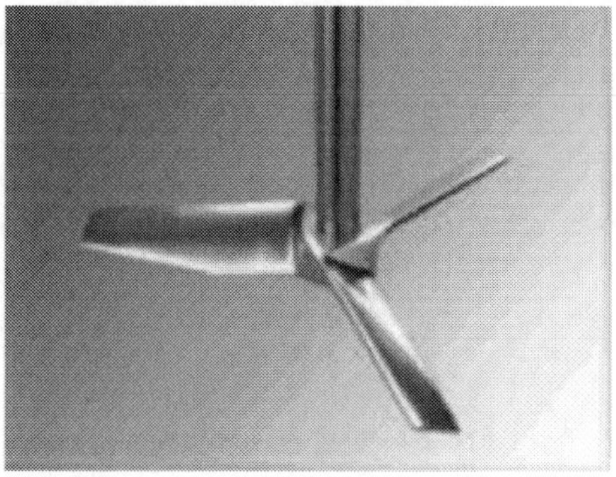

Figure 1: A310 Impeller

Figure 2: A200 Impeller

423c Coalescence Factor: A Dimensionless Group to Predict the Coalescence of Dynamic Foams

Weiguo Xie, Stephen Neethling, Jan Cilliers

Much research has investigated the stability of foams, however, predicting dynamic foam stability remain elusive. Building on the law of conservation of momentum and using dimension analysis (Buckingham PI Theory), a dimensionless group, coalescence factor C_f, was defined to indicate the onset of coalescence. Coalescence factor describes the balance between dynamic surface tension and the applied deformation force.

Experiments show that changes of air flow-rates, orifice diameter, surfactant concentration, and type of surfactant (SDS and Teepol) affect the coalescence factor. Figure 1 shows an example of the coalescence factor for different orifice diameters (5.0cm, 1.5cm, and 0.3cm) at the CMC (critical micelle concentration) of Teepol with varying air flowrate. Close agreement was found between the onset of coalescence, determined by image analysis and ERT and a coalescence factor greater than unity. The relationship between coalescence factor and N_c, the number of coalescence events, also showed a critical point at coalescence factor >1.0. Through various experiments, it has been shown that coalescence factor can predict the coalescence of dynamic foams.

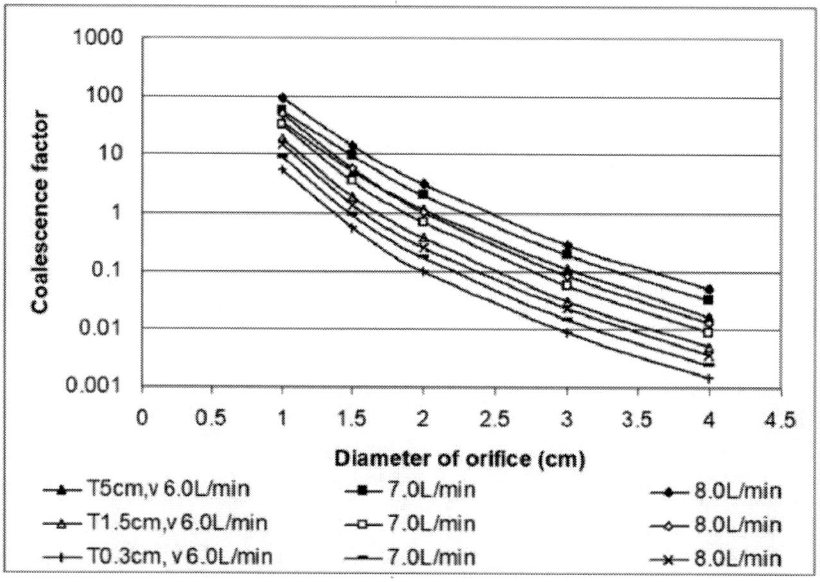

Figure 1: Coalescence factor for different diameter of orifice at CMC (critical micelle concentration) of Teepol using three different thickness of orifice (5.0cm, 1.5cm, and 0.3cm) with varying air flowrate.

423d CFD Validation in Gas-Liquid Systems Using Flow Flowing Sensors

Christopher Tyler

There is a constant need to validate computational fluid dynamics models against real data to test and improve methods. Many means of validating CFD against experiments have been proposed and utilized, but few allow testing in industrial scale operations. By using flow-following sensors, neutrally buoyant probes approximately 30 mm in diameter, we have been able to measure position, velocity, and scalar fields at industrial scale in gas/liquid systems, with sampling periods of multiple days, on actual systems, during normal operations. For sparged tanks, with and without agitation, we measured the position, velocity, and scalar fields with a few probes for a few days of operation. We have also simulated these systems using a Lattice Boltzmann Method, with hundreds of probes, collecting data for many probes over shorter periods of time, to produce datasets of similar size. Appealing to the principle of ergodicity, we compare the distributions of probe positions, velocities, and probe-circulation times. We find that the simulated and experimental position, velocity, and circulation time distributions match with an R-squared of 0.90 or greater. Analyzing the datasets as distributions also provides a simple means of comparing different operating conditions, such as operation with and without agitation. It similarly provides means of comparing differences from model formalism, such as different choices of breakup and coalescence models in sparged systems.

423f On the Just Suspended Speed in Widely Graded Slurries

Jason Giacomelli, Richard Grenville, Harry Van den Akker

Typical correlations to calculate an impeller's just suspended speed are based upon experiments with slurries with narrow particle size distributions and concentrations smaller than 20-30%v/v in which suspension is due to turbulent eddies in the inertial subrange. In the minerals processing industry, however, particles are often widely graded with a significant portion of them (≈>50%) being smaller than 75 μm. While at elevated concentrations (>20-30% v/v), these types of slurries can become viscous and even non-Newtonian, we do not have an applicable correlation. New suspension experiments are presented for three model widely graded slurries over concentrations of $1 < x_v < 50\%$v/v. As the concentration increases, the suspension mechanism changes from inertial subrange turbulence to the viscous regime. We present a preliminary correlation for the viscous suspension regime, for which one is not currently available. This correlation exploits a suspension viscosity due to the presence of the fines.

AUTHOR INDEX